Ernst Schering Research Foundation Workshop
Supplement 6
Testis, Epididymis and Technologies in the Year 2000

Springer-Verlag Berlin Heidelberg GmbH

Ernst Schering Research Foundation Workshop
Supplement 6

Testis, Epididymis and Technologies in the Year 2000

11th European Workshop on Molecular and Cellular Endocrinology of the Testis

B. Jégou, C. Pineau, J. Saez
Editors

With 29 Figures

Springer

Series Editors: G. Stock and M. Lessl

ISSN 0947-6075
ISBN 978-3-662-04052-2

CIP data applied for

Die Deutsche Bibliothek – CIP-Einheitsaufnahme
Testis, epididymis and technologies in the year 2000 / 11th European Workshop on Molecular and Cellular Endocrinology of the Testis. Ernst Schering Research Foundation. Bernard Jégou ... ed.
(Ernst Schering Research Foundation Workshop : Supplement ; 6)

ISBN 978-3-662-04052-2 ISBN 978-3-662-04050-8 (eBook)
DOI 10.1007/978-3-662-04050-8

Originally published by Springer-Verlag Berlin Heidelberg New York in 2000
Softcover reprint of the hardcover 1st edition 2000

Typesetting: Data conversion by Springer-Verlag

SPIN:10765725 13/3134–5 4 3 2 1 0 – Printed on acid-free paper

Preface

The first " European Workshop on Molecular and Cellular Endocrinology of the Testis" was held in Geilo, Norway, in April 1980. Since then, this most prominent forum for European and international research in the field of male reproduction has taken place every other year in Holland, France, Italy, England, Sweden/Finland (Åland Islands), Germany, Belgium and Italy, respectively.

Twenty years after its foundation, that is during the mythical year 2000 (13–17 May), we have had the privilege of organizing the 11th Workshop in Saint-Malo, Brittany, France. The main lectures of this Workshop have been compiled in this volume. A number of chapters present the most novel research on testicular and epididymal functions or on more general fields of hormone action and molecular cell biology as it is now a tradition. However, exceptionally, the book also contains several chapters dealing with the " Approaches and Tools in the Third Millenium ". The unusual inclusion of Technologies as such in the year 2000 edition of the Workshop was an absolute necessity, as these technologies are revolutionizing the fields of biology and medicine and, in many instances, how to do research. Therefore, we felt that it was essential for our scientific community to be informed of the very latest technical developments and their potential for future progress.

This book was edited thanks to the generous support of the Ernst Schering Research Foundation. We also want to thank the distinguished scientists from three different continents, who have provided us with their high quality manuscripts well before the Workshop, and Dr. U.-F. Habenicht, Dr. M. Lessl and Ms W. McHugh for making this publication possible in such a short time.

We also acknowledge the support of the International Scientific Committee of the European Testis Workshop, as well as of the members of the National and Local Organizing Committees as listed below.

B. Jégou, C. Pineau and J. Saez

Permanent Scientific Committee of the European Workshop on Molecular and Cellular Endocrinology of the Testis
Brian A. Cooke, London; Vidar Hansson, Oslo; Ilpo Huhtaniemi, Turku; Bernard Jégou, Rennes; Eberhard Nieschlag, Münster; Focko Rommerts, Rotterdam; Olle Söder, Stockholm; Mario Stefanini, Rome; Guido Verhoeven, Leuven.

National Organizing Committee of the 11th European Workshop on Molecular and Cellular Endocrinology of the Testis
Philippe Berta, Montpellier; Philippe Bouchard, Paris; Serge Carreau, Caen; Philippe Durand, Lyon; Marc Fellous, Paris; Georges Guellaën, Créteil; Florian Guillou, Nouzilly; Bernard Jégou, Rennes; Pierre Jouannet, Paris; Florence Le Gac, Rennes; Edwin Milgrom, Bicêtre; Jean-Yves Picard, Montrouge; José Saez, Lyon; Paolo Sassone-Corsi, Illkirch.

Local Organizing Committee of 11th European Workshop on Molecular and Cellular Endocrinology of the Testis in Rennes
Bernard Jégou, Charles Pineau, Noureddine Boujrad, Alexis Fostier, Jean-Jacques Lareyre, Florence Le Gac, Michel Samson.

Table of Contents

1 Analysis of the Meiotic Transcriptome in Genetically Distinct Budding Yeasts Using High Density Oligonucleotide Arrays
M. Priming, R.M. Williams, E. A. Winzeler, G. G. Tevzadze, A. R. Conway, S.Y. Hwang, R. W. Davis, R.E.Esposito . . . 1

2 Cell and Antibody Engineering: New Tools for Antigen and Therapeutic Antibody Discovery
J.P. Mather, L. Bald, P. E. Roberts, R.-h. Li, J-Ph. Stephan 9

3 The Wonderful World of Proteomics. The Key Role of MDS Protana A/S
K.L. Bennett, E.Mørtz 27

4 Cyclin-dependent Kinases and their Chemical Inhibitors: New Targets and Tools for the Study of the Testis?
L. Meijer, M. Leost, S. Leclerc, M. Garnier 39

5 Genetic Manipulation and Transplantation of Male Germ Cells
T. Guillaudeux, C. Celebi, P. Auvray, B. Jégou 59

6 Germ Cell Culture, Genetic Manipulation and Transplantation
S. Schlatt, A.G. Schepers, V. von Schönfeldt 69

7 Regulatory Mechanisms in Mammalian Spermatogenesis
D.M. de Kretser, T. Meehan, M.K. O'Bryan, N.G. Wreford, R.I. McLachlan, K.L. Loveland 87

8 Routes of Transcriptional Activation in the Testis: CREM and its Coactivator ACT
D. de Cesare, G.M. Fimia, A. Morlon, P. Sassone-Corsi . . 107

9 Apoptosis- Searching for the Central Executioner
E. Daugas, G. Kroemer 129

10 Germ Cell Apoptosis
M.E. Embree, K. Boekelheide 141

11 Recruitment of p160 Coactivators to Androgen Receptors
M. Parker and Ch. Bevan 165

12 Interaction of Androgens and Oestrogens in Development of the Testis and Male Reproductive Tract
R.M. Sharpe, C. McKinnell, N. Atanassova, K. Williams, K.J. Turner, PTK Saunders, M. Walker, M.R. Millar, J.S Fisher 173

13 Sertoli Cell Proteins in Testicular Paracriny
D.D. Mruk, C.Y. Cheng 197

14 The Coming Age of the Epididymis
B. Robaire, P. Syntin, K. Jervis 229

Subject Index 263

Previous Volumes Published in This Series 267

List of Editors and Contributors

Editors

B. Jégou
GERM-INSERM U. 435, Université de Rennes, Campus de Beaulieu,
35042 Rennes, Bretagne, France

C. Pineau
GERM-INSERM U. 435, Université de Rennes, Campus de Beaulieu,
35042 Rennes, Bretagne, France

J. Saez
INSERM U. 369, Faculté de médicine Lyon-RTH Laennec, rue Guillaume
Paradin, 69372 Lyon, Cedex 08, France

Contributors

N. Atanassova
MRC Reproductive Biology Unit, Centre for Reproductive Biology,
37 Chalmers Street, Edinburgh EH3 9ET, Scotland, UK

P. Auvray
GERM-INSERM U. 435, Université de Rennes, Campus de Beaulieu,
35042 Rennes, Bretagne, France

L. Bald
Raven Biotechnologies, Inc. 305 Old County Road, San Carlos, CA 94070,
USA

K.L. Bennett
MDS Protana A/S, Staermosegaardsvej 16, DK-5230 Odense M, Denmark

Ch. Bevan
Department of Cancer Medicine, Division of Medicine, Imperial College of Science, Technology and Medicine, MRC Cyclotron Building, Hammersmith Hospital, Du Cane Road, London W12 ONN, UK

K. Boekelheide
Department of Pathology and Laboratory Medicine, Brown University, Providence, RI 02912, USA

C. Celebi
GERM-INSERM U. 435, Université de Rennes I, Campus de Beaulieu, 35042 Rennes, Bretagne, France

D. de Cesare
Institut de Génétique et de Biologie Moléculaire et Cellulaire, CNRS-INSERM, Université Louis Pasteur, BP 163, 67404 Illkirch-Strasbourg

C.Y. Cheng
Population Council, Center for Biomedical Research, 1230 York Avenue, New York, New York 10021, USA

A.R. Conway
Silicon Genetics, 935 Washington Street, San Carlos, CA, 94070, USA

E. Daugas
Centre National de la Recherche Scientifique CNRS, 1599, Institut Gustave Roussy, 39 rue Camille-Desmoulins, 94805 Villejuif, France

R.W. Davis
Stanford University Medical School, Department of Biochemistry, Stanford, CA 94305—5307, USA

M.E. Embree
Department of Pathology and Laboratory Medicine, Brown University, Providence, RI 02912, USA

R.E. Esposito
The University of Chicago, Department of Molecular Genetics and Cell Biology, 920 E, 58th Street, Chicago, IL 60637, USA

G.M. Fimia
Institut de Génétique et de Biologie Moléculaire et Cellulaire, CNRS-INSERM, Université Louis Pasteur, BP 163, 67404 Strasbourg

J.S. Fisher
MRC Reproductive Biology Unit, Centre for Reproductive Biology, 37 Chalmers Street, Edinburgh EH3 9ET, Scotland, UK

M. Garnier
Station Biologique de Roscoff, CNRS UPR 9042, BP 74, 29682 Roscoff, Bretagne, France

T. Guillaudeux
GERM-INSERM U. 435, Université de Rennes, Campus de Beaulieu, 35042 Rennes, Bretagne, France

S.Y. Hwang
Stanford University Medical School, Department of Biochemistry, Stanford, CA 94305—5307, USA

K. Jervis
Department of Pharmacology and Therapeutics, McGill University, 3655 Promenade Sir-William-Osler, Montréal, Québec, Canada, H3G 1Y6

D.M. de Kretser
Monash Institute of Reproduction and Development, Monash Medical Center, 246 Clayton Road, Clayton, Victoria, Australia

G. Kroemer
Centre National de la Recherche Scientifique CNRS, 1599, Institut Gustave Roussy, 39 rue Camille-Desmoulins, 94805 Villejuif, France

S. Leclerc
Station Biologique de Roscoff, CNRS UPR 9042, BP 74, 29682 Roscoff, Bretagne, France

M. Leost
Station Biologique de Roscoff, CNRS UPR 9042, BP 74,
29682 Roscoff, Bretagne, France

R.-h. Li
Raven Biotechnologies, Inc. 305 Old County Road, San Carlos, CA 94070,
USA

K.L. Loveland
Monash Institute of Reprodution and Development, Monash Medical Centre,
246 Clayton Road, Clayton, Vic 3168, Australia

J.P. Mather
Raven Biotechnologies, Inc. 305 Old County Road, San Carlos, CA 94070,
USA

R.I. McLachlan
Prince Henry's Institute of Medical Research, Melbourne, Victoria, Australia

C. McKinnell
MRC Reproductive Biology Unit, Centre for Reproductive Biology,
37 Chalmers Street, Edinburgh EH3 9ET, Scotland, UK

T. Meehan
Monash Institute of Reprodution and Development, Monash Medical Centre,
246 Clayton Road, Clayton, Vic 3168, Australia

L. Meijer
Station Biologique de Roscoff, CNRS UPR 9042, BP 74,
29682 Roscoff, Bretagne, France

M.R. Millar
MRC Reproductive Biology Unit, Centre for Reproductive Biology,
37 Chalmers Street, Edinburgh EH3 9ET, Scotland, UK

A. Morlon
Institut de Génétique et de Biologie Moléculaire et Cellulaire,
CNRS-INSERM, Université Louis Pasteur, BP 163, 67404 Illkirch-Strasbourg

E. Mortz
MDS Protana A/S, Staermosegaardsvej 16, DK-5230 Odense M, Denmark

D.D. Mruk
Population Council, Center for Biomedical Research, 1230 York avenue, New York, New York 10021, USA

M.K. O'Bryan
Monash Institute of Reprodution and Development, Monash Medical Centre, 246 Clayton Road, Clayton, Vic 3168, Australia

M. Parker
Molecular Endocrinology, Laboratory, Imperial Cancer Research Fund, 44 Lincoln's Inn Fields, London WC2 A 3PX, UK

M. Priming
Institut de Génétique Humaine, 141 rue de la Cardonille, 34396 Montpellier, France

B. Robaire
Department of Pharmacology and Therapeutics, McGill University, 3655 Promenade Sir-William-Osler, Montréal, Québec, Canada, H3G 1Y6

P.E. Roberts
Raven Biotechnologies, Inc. 305 Old County Road, San Carlos, CA 94070, USA

P. Sassone-Corsi
Institut de Génétique et de Biologie Moléculare et Cellulaire, CNRS-INSERM, Université Louis Pasteur, BP 163, 67404 Illkirch- Strasbourg, France

P.T.K. Saunders
MRC Reproductive Biology Unit, Centre for Reproductive Biology, 37 Chalmers Street, Edinburgh EH3 9ET, Scotland, UK

A.G. Schepers
Institute of Reproductive Medicine,University of Münster, Domagkstraße 11, 48149 Münster, Germany

St. Schlatt
Institute of Reproductive Medicine, University of Münster, Domagkstraße 11, 48149 Münster, Germany

R.M. Sharpe
MRC Reproductive Biology Unit, Centre for Reproductive Biology,
37 Chalmers Street, Edinburgh EH3 9ET, Scotland, UK

J.-Ph. Stephan
Genentech, Inc., I DNA Way, South San Francisco, CA 94080, USA

P. Syntin
Department of Pharmacology and Therapeutics, McGill University,
3655 Promenade Sir-William-Osler, Montréal, Québec, Canada, H3G 1Y6

G.G. Tevzadze
The University of Chicago, Department of Molecular Genetics and Cell Biology, 920 E, 58th Street, Chicago, IL 60637, USA

V. von Schönfeld
Institute of Reproductive Medicine of the University, Domagkstraße 11,
48149 Münster, Germany

K.J. Turner
MRC Reproductive Biology Unit, Centre for Reproductive Biology,
37 Chalmers Street, Edinburgh EH3 9ET, Scotland, UK

M. Walker
MRC Reproductive Biology Unit, Centre for Reproductive Biology,
37 Chalmers Street, Edinburgh EH3 9ET, Scotland, UK

K. Williams
MRC Reproductive Biology Unit, Centre for Reproductive Biology,
37 Chalmers Street, Edinburgh EH3 9ET, Scotland, UK

R.M. Williams
The University of Chicago, Department of Molecular Genetics and Cell Biology, 920 E, 58th Street, Chicago, IL 60637, USA

E.A. Winzeler
Stanford University Medical School, Department of Biochemistry,
Stanford, CA 94305—5307, USA

N.G. Wreford
Department of Anatomy, Monash University, Melbourne, Victoria, Australia

1 Analysis of the Meiotic Transcriptome in Genetically Distinct Budding Yeasts Using High Density Oligonucleotide Arrays

M. Primig, R. M. Williams, E. A. Winzeler, G. G. Tevzadze,
A. R. Conway, S. Y. Hwang, R. W. Davis, R. E. Esposito

1.1 High Density Oligonucleotide and cDNA Array Technologies 1
1.2 Meiotic Development and Gene Expression in the Budding Yeast .. 4
1.3 Conclusions 5
References 6

1.1 High Density Oligonucleotide and cDNA Array Technologies

The last three years have witnessed a massive production of whole-genome expression data in the yeast field using two different types of gene arrays: commercially available high-density oligonucleotide arrays (GeneChips) and PCR-based gene arrays. In the case of GeneChips every yeast gene is represented by 20 oligonucleotides (each being a 25-mer) synthesized *in situ* onto a glass plate (which is then inserted into a cartridge for experimental manipulation; Fig. 1, panel b, top left image). *Poly* A^+ RNA (e.g. prepared from cells at various stages of spore development; Fig. 1, panel a) is reverse transcribed into cDNA, labeled with a fluorophor and hybridized to the GeneChip. The fluorescence signal intensities of each set of 20 oligonucleotides are directly propor-

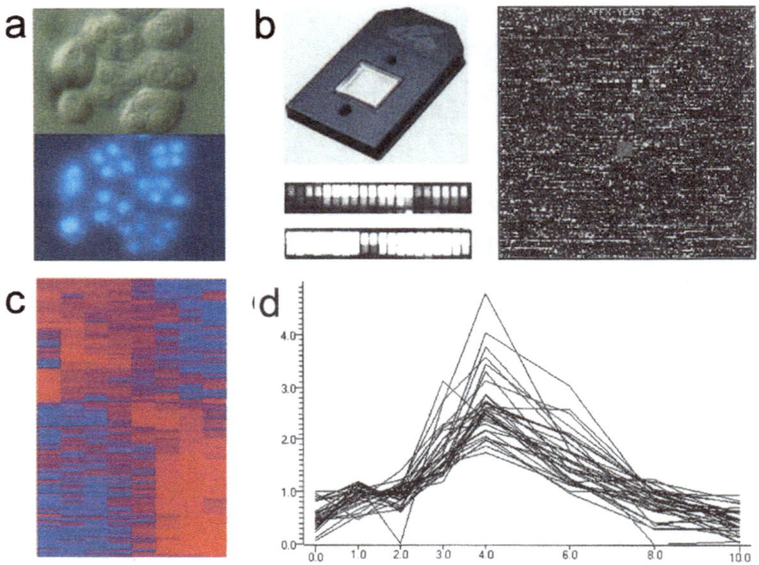

Fig. 1a–d. A GeneChip-based analysis of the meiotic transcriptome in yeast. Panel (a) shows a Nomarski view (top half) and Hoechst-stained nuclei using UV-light microscopy (bottom half) of mature asci formed after 12 hours of spore development in the yeast strain SK1. Panel (b) summarizes the image of a GeneChip cartridge (top left), the result of a fluorescence scan (right) and individual hybridization results indicating correct ratios between perfect match and mismatch (middle left) and a case of clear cross-hybridization (bottom left). Panel (c) shows a bar diagram of 130 selected meiotic expression patterns as identified in SK1 (ordered over initial time of induction and then clustered on the basis of overall similarity). Each column corresponds to a time-point (t=0, 1, 2, 3, 4, 6, 8 and 10 hours of sporulation), each line represents a gene. Red and blue indicate high and low levels of expression, respectively. Panel (d) displays the graphical view of 33 selected meiotically induced genes whose expression patterns are very similar (correlation coefficient 0.9). Fluorescence intensities that are directly proportional to mRNA concentrations are plotted versus hours of sporulation in SK1

tional to the mRNA concentration in the sample (Fig. 1, panel b, image of a GeneChip hybridization pattern). To eliminate the problem of cross hybridization a set of wild-type oligonucleotides (perfect match) are compared to a set of oligonucleotides containing a point mutation (mismatch) that destabilizes the DNA-DNA interaction. Two examples of a correct hybridization pattern (panel b, middle left) and a clear case of cross-hybridization (panel b, bottom left) are shown in Fig. 1.

cDNA-based arrays can be purchased from a rapidly growing number of suppliers or manufactured by individual research laboratories (Bowtell, 1999). Various companies offer the necessary equipment to make and scan arrays. In the case of gene arrays a PCR fragment covering the entire open reading frame represents each locus. This DNA fragment is spotted onto a glass plate in duplicate. A hybridization mix contains two cDNA probes obtained from different strains (e.g. wild type versus mutant) or from cells cultured under distinct growth conditions; these probes are labeled with two different fluorophores. The result produced after a hybridization reaction is the ratio of the fluorescence intensities obtained with the control sample and the sample to be analysed.

Both array-based approaches yield data at an unprecedented scale and require specialized software for statistical data analysis (clustering; see Eisen et al., 1998) and graphical presentation (e.g. Matlab and GeneSpring; see www.sigenetics.com; Fig. 1, panels c and d). Whole-genome expression studies have yielded important information about the state of transcription of nearly all yeast genes under various growth conditions and during meiotic development (Cho et al., 1998; Chu et al., 1998; DeRisi et al., 1997; Holstege et al., 1998; Jelinsky and Samson, 1999; Spellman et al., 1998). Useful information about gene arrays can be found at http://www.gene-chips.com and in a recent comprehensive review of all aspects of array usage and manufacturing (Phimister, 1999).

1.2 Meiotic Development and Gene Expression in the Budding Yeast

Saccharomyces cerevisiae

Gametogenesis in budding yeast, which consists of meiosis and spore development, has been used to identify many genes required for meiotic development (Kupiec et al., 1997). Meiotic landmark events including initiation of recombination, formation of the synaptonemal complex and checkpoint controls regulating M-phase progression are conserved from yeast to mammals, often utilizing homologous gene functions (Edelman et al., 1999; Freire et al., 1998; McKim and Hayashi-Hagihara, 1998; Romanienko and Camerini-Otero, 1999). Hence, understanding meiosis in yeast should facilitate the identification of genes important for gametogenesis in multi-cellular organisms.

In this study two genetically distinct strains (SK1 and W303) are examined using high-density oligonucleotide arrays to identify the complete profile of meiotically regulated transcripts in yeast. Initiation and completion of spore development in yeast requires the *MAT***a**/*MAT*α locus and the absence of both nitrogen and fermentable carbon sources. Classical approaches like the isolation of sporulation-deficient mutants, cloning of differentially expressed genes, and analysis of meiotic phenotypes of cell cycle mutants have identified ~200 genes required for meiotic development (Kupiec et al., 1997). Among these, the expression patterns of ~40 sporulation genes were analyzed in detail and shown to fall into at least five expression groups: very early, early, middle, mid-late and late (Mitchell, 1994). Three transcription factors are currently known to participate in the transcriptional regulation of these genes, Ume6p/Ime1p (Anderson et al., 1995; Steber and Esposito, 1995; Strich et al., 1994; Sweet et al., 1997), Abf1p (Gailus-Durner et al., 1996; Ozsarac et al., 1997) and Ndt80p, which is present exclusively in sporulating cells (Chu and Herskowitz, 1998; Hepworth et al., 1998). Several hundred additional transcriptionally induced genes, revealed by recent cDNA-array expression analysis, generally support these prior group designations (Chu et al., 1998).

The study presented here (i) defines a common set of ~ 900 meiotically regulated genes in both strains, (ii) identifies several hundred genes displaying strain-specific meiotic expression patterns, which may account for their different sporulation properties, and (iii) distinguishes genes involved in the response to nutrient deprivation from those specifically required for the landmark events of sporulation. The set of meiotically expressed genes identified contains the majority of known sporulation genes and ~ 400 additional loci not found in a recent independent whole genome expression study (Chu et al., 1998).

1.3 Conclusions

By integrating data from two independent sporulation expression analyses we have compiled a list of ~ 900 strain-independently expressed genes (see Fig. 1, panels c and d), which potentially have the greatest relevance to sporulation in yeast. In addition, we have verified that the majority of these genes are differentially expressed only during sporulation and not during starvation; hence they are expressed in a manner specific to this developmental program. In comparison to the previous cDNA microarray based study of sporulation performed by Chu *et al.* (1998) we find approximately 400 genes to be up regulated that were not detected in the previous study. Some of the discrepancies can be accounted for by the different approaches taken in the preparation of the biological samples and the analytical techniques used to assess the data. It is also possible that the low synchrony of sporulation induced in the experiment by Chu *et al.* might have masked the appearance of some genes. These issues aside, the main regions of agreement between the experiments cover almost all the previously known sporulation genes and clearly implicate several hundred new genes as potentially important for meiotic development in yeast. A clear conclusion from the strain comparison is that the timing of the transcriptional events and the timing of the morphological changes during sporulation coincide very closely, that is to say the fast SK1 strain induces meiotic genes consistently earlier than the slow W303 strain. Preliminary results from the EUROFAN genomic gene deletion project indicate that the pool of induced genes identified in this work indeed contains many essential sporulation genes (A. Nicholas, personal communication). This suggests that genes important to the process are generally differentially expressed; however the observed effects are sometimes small, and could be confused with the slight fluctuations measured in the expression of many genes from sample to sample. The use of bar-coded deletion strains (Winzeler and Davis, 1999), now constructed for almost every annotated ORF in the yeast genome, as well as other gene-inactivation strategies (Ross-Macdonald et al., 1999), offers an opportunity to evaluate the actual biological contribution each gene makes to a particular process.

References

Anderson S, Steber C, Esposito R, Coleman J (1995) UME6, a negative regulator of meiosis in Saccharomyces cerevisiae, contains a C-terminal Zn2Cys6 binuclear cluster that binds the URS1 DNA sequence in a zinc-dependent manner. Protein Sci 4:1832–1843

Bowtell D (1999) Options available – from start to finish – for obtaining expression data by microarray. Nature Genetics 21:25–32

Cho RJ, Campbell MJ, Winzeler EA, Steinmetz L, Conway A, Wodicka L, Wolfsberg TG, Gabrielian AE, Landsman D, Lockhart DJ, Davis RW (1998) A genome-wide transcriptional analysis of the mitotic cell cycle. Molecular Cell 2:65–73

Chu S, DeRisi J, Eisen M, Mulholland J, Botstein D, Brown P, Herskowitz I (1998) The transcriptional program of sporulation in budding yeast. Science 282:699–705

Chu S, Herskowitz I (1998) Gametogenesis in yeast is regulated by a transcriptional cascade dependent on Ndt80. Cell 1:685–696

DeRisi JL, Iyer VR, Brown PO (1997) Exploring the metabolic and genetic control of gene expression on a genomic scale. Science 278:680–686

Edelman W, Cohen P, Kneitz B, Winand N, Lia M, Heyer J, Kolodner R, Pollard J, Kucherlapati R (1999) Mammalian MutS homologue 5 is required for chromosome pairing in meiosis. Nature Genetics 21:123–127

Eisen M, Spellman P, Brown P, Botstein D (1998) Cluster analysis and display of genome-wide expression patterns. Proc Natl Acad Sci USA 95:14863–14868

Freire R, Murguia J, R, Tarsounas M, Lowndes N F, Moens PB, Jackson SP (1998) Human and mouse homologs of Schizosaccharomyces pombe rad1+ and Saccharomyces cerevisiae RAD17: linkage to checkpoint control and mammalian meiosis. Genes & Dev 12:2560–2573

Gailus-Durner V, Xie J, Chintamaneni C, Vershon AK (1996) Participation of the yeast activator Abf1 in meiosis-specific expression of the HOP1 gene. Molecular & Cellular Biology 16:2777–86

Hepworth SR, Friesen H, Segall J (1998) NDT80 and the meiotic recombination checkpoint regulate expression of middle sporulation-specific genes in Saccharomyces cerevisae. Mol Cell Biol 18:5750–5761

Holstege F, Jennings E, Wyrick J, Lee T, Hengartner C, Greeen M, Golub T, Lander E, Young R (1998) Dissecting the regulatory circuitry of a eucaryotic genome. Cell 95:717–728

Jelinsky S, Samson L (1999) Global response of Saccharomyces cerevisiae to an alkylating reagent. Proc Natl Acad Sci USA 96:1486–1491

Kupiec M, Byers B, Esposito R, Mitchell A (1997) Meiosis and Sporulation in Saccharomyces cerevisiae. Cold Spring Harbour Laboratory Press, 889–1036

McKim K, Hayashi-Hagihara A (1998) mei-W68 in Drosophila melanogaster encodes a Spo11 homolog: evidence that the mechanism for initiating meiotic recombination is conserved. Genes & Dev 12:2932–2942

Mitchell AP (1994) Control of meiotic gene expression in Saccharomyces cerevisiae. Microbiological Reviews 58:56–70

Ozsarac N, Straffon MJ, Dalton HE, Dawes IW (1997) Regulation of gene expression during meiosis in Saccharomyces cerevisiae: SPR3 is controlled by both ABFI and a new sporulation control element. Molecular & Cellular Biology 17:1152–9

Phimister B (1999) Going global. Nature Genetics 21

Romanienko P, Camerini-Otero R (1999) Cloning, characterization, and localization of mouse and human SPO11. Genomics 61:159–159

Ross-Macdonald P, Coelho P, Roemer T, Agarwal S, Kumar A, Jansen R, Cheung K, Sheehan A, Symoniatis D, Umansky L, Heidtman M, Nelson F, Iwasaki H, Hagers K, Gerstein M, Miller P, Roeder G, Snyder M (1999) Large-scale analysis of the yeast genome by transposon tagging and gene disruption. Nature 402:413–418

Spellman PT, Sherlock G, Zhang MQ, Iyer VR, Anders K, Eisen MB, Brown PO, Botstein D, Futcher B (1998) Comprehensive identification of cell cycle-regulated genes of the yeast Saccharomyces cerevisiae by microarray hybridization. Molecular Biology of the Cell 9:3273–97

Steber CM, Esposito RE (1995) UME6 is a central component of a developmental regulatory switch controlling meiosis-specific gene expression. Proceedings of the National Academy of Sciences of the United States of America 92:12490–4

Strich R, Surosky RT, Steber C, Dubois E, Messenguy F, Esposito RE (1994) UME6 is a key regulator of nitrogen repression and meiotic development. Genes & Development 8:796–810

Sweet D, Jang Y, Sancar G (1997) Role of UME6 in transcriptional regulation of a DNA repair gene in Saccharomyces cerevisiae. Mol Cell Biol 17:6223–6235

Winzeler E, Davis R (1999) Functional characterization of the S. cerevisiae genome by gene deletion and parallel analysis. Science 285:901–906

2 Cell and Antibody Engineering: New Tools for Antigen and Therapeutic Antibody Discovery

J. P. Mather, L. Bald, P. E. Roberts, R.-h. Li, J.-P. Stephan

2.1 Introduction: Bioscience in the 21st Century 9
2.2 Cell Culture: A New Era of Novel Cell Lines 10
2.3 Cell Surface Proteins: Important Regulators of Survival, Growth and Development 16
2.4 Raising Antibodies to Cell Surface Proteins. A New Approach to an Established Technology Yields Unique Set of Antibodies 17
2.5 An Example: Immunization with Intact Rat Fetal Pancreatic Cells 19
2.6 Novel Cell Lines and Panels of Mabs to Cell Surface 21
References 23

2.1 Introduction: Bioscience in the 21st Century

The inception of genetic engineering, the complete sequencing of human, animal and plant genomes, and the information revolution in the last quarter of the 20th century will forever change the way in which biological and medical science is carried out in the laboratory. These technologies have spawned new tools and techniques that have created huge amounts of data. However, with the increased use of these tools, there is the growing realization that the gene sequences, alone, will not be sufficient to allow us to understand the organisms that they create. With the entire sequence of the human genome likely to be available in the near future, one of the challenges for the next century will be to identify the subset of known genes that are important in regulating the development of a specific organ. Once these genes are identified, one would wish to have biologically relevant model systems to rapidly assess the function of individual genes.

The mammalian cell is the basic unit that performs most functions in the body producing and responding to hormones, fighting foreign or diseased cells, healing wounds, and providing ongoing replacement of cells to repair and maintain organs. The ideal would be to acquire the ability to maintain individual cells, or groups of defined cell types, in culture in a manner that allows the cells to continue to function as they do in the body. This would allow us to use these cultures as model systems to understand normal development, degenerative diseases, and perhaps eventually, to stimulate tissue repair or regeneration at will.

2.2 Cell Culture: A New Era of Novel Cell Lines

Cell culture, the ability to maintain isolated mammalian cells outside the body, has existed, in some form, for 90 years (Mather and Roberts, 1998). Most of these cells are cultured in a nutrient medium containing salts, sugars, and amino acids, supplemented with serum (the liquid remaining after clotting whole blood) or some other complex, undefined mixture of proteins. In the last 40 years many cell lines have been established and grown in serum-containing medium. In order to grow in these conditions most cells had to be chemically or virally transformed, or established from tumor tissue. In the majority of these cases, only a few cells from the original culture were able to adapt to the culture conditions and continue to grow.

In the last decade, significant progress has been made in understanding the requirements for maintaining normal cells *in vitro*, and in understanding what caused the loss of normal function in so many of the cell lines established in the past. This has allowed the culture of both human embryonic stem cell lines (Thomson *et al.*, 1998) and developmentally arrested tissue or organ specific stem cells from the hematopoetic system, nervous system (Li *et al.*, 1996a; Loo *et al.*, 1987), and other organs. If one thinks of all of the different cell types that exist in the animal in different tissues, as well as the different stages of development that many of these cell types go through, it is clear that many hundreds of distinct cell phenotypes exist *in vivo*. Most of these have yet to be maintained in culture for *in vitro* studies and many not yet described and characterized even *in vivo*.

Establishing cell lines, from the initial primary culture, in serum-free media (supplemented with growth factors, vitamins, attachment factors and hormones) allows us to create novel cell lines that have unique properties compared to the vast majority of cell lines established in the traditional manner in serum-containing media.

- Defined serum-free media select for the growth of a single cell type.
- Many cell lines established in this fashion will not grow in serum and are therefore, by definition, a different phenotype than any of the cell lines commonly grown in serum.
- These media can be devised to "developmentally arrest" fetal cells in specific stage of development allowing the establishment of progenitor cell lines.
- Cells grown in serum-free medium are more genetically and phenotypically stable than those grown in serum-containing medium.
- Normal rodent cells become immortal cell lines without transformation. Normal human cells have extended lifespans.

We have taken serum-free and defined culture into new areas by establishing and continuously maintaining a number of normal dividing cell types in serum-free media.

Growing cell lines without serum is a relatively new technology and includes various approaches from short-term maintenance in serum-free media to establishing and passaging cells in defined conditions. We describe below four different types of cells and their ability to grow in serum vs. serum-free media (Murakami *et al.*, 1985).

2.2.1 Class I: Cell Lines That Were Established in Media That Contain Serum and Continue to Be Routinely Grown with Serum Supplemented Medium

These cells may be put into a serum free culture, with appropriate hormone supplements, for specific purposes, e.g. recombinant protein production or for a short-term experiment. This serum-free technology follows from the work on replacing serum with growth factors that came out of Dr. Gordon Sato's laboratory in the period 1976–1985 (Barnes and Sato, 1980; Bottenstein *et al.*, 1979). Of the many types of cells grown in serum, only a few are used in this manner (Mather and Barnes, 1998; Murakami *et al.*, 1985).

Class I cells are useful for specific purposes, including recombinant protein production, studying the short-term effects of a single hormone, and understanding the role serum plays in promoting cell growth. However, their usefulness for product discovery and understanding normal cell function is limited. The cells' properties have often changed considerably from what they were in the body. In addition, even when the cell is removed from a medium that contains bovine serum, numerous serum proteins adhere to the cell surface and make identification of cell-produced proteins more difficult.

2.2.2 Class II: Cells for Which Serum Supplementation Is Inadequate to Support Growth

There are a number of cell types that are not harmed by serum but still will not grow even in the presence of 10–20% serum because something is needed that is not present in the serum in sufficient amounts. These include a number of neuronal cell lines that require neurotrophic factors and other cell types (e.g. Schwann cells, endothelial cells, and some neuroepithelial cells) that require tissue- or cell-specific growth factors (Frederiksen *et al.*, 1988; Morrissey *et al.*, 1995). Many of these paracrine factors are only present in high enough levels in the tissues where they are made and, unlike hormones, are never released into the general circulation. Media that support the growth of these cells, e.g. media that contain serum supplemented with other growth factors, are not selective. In addition to the cell of choice, these media will support the growth of any cell that will grow in serum (e.g. fibroblasts). Several dozen labs are working with these types of cells and several cell lines exist which require growth factors in addition to serum.

If a cell line is started from primary culture in such a medium, it is difficult to obtain a pure cell line because many other cells (e.g. fibroblasts) from the original tissue will also grow in serum-containing medium. We prefer to establish this class of cells directly from primary culture in a defined serum-free medium. We can then obtain a pure, functional cell line consisting of only the cell type of interest without potential changes, and chromosomal instability, that might be introduced by growth in serum. Thus the results from working with these cell lines are more repeatable and predictive of the behavior of the cell *in vivo* (cells *in vivo* are seldom exposed to serum).

2.2.3 Class III: Cells That Are Differentiated or Growth Arrested by Serum

Serum contains differentiation factors that will induce some types of cells to differentiated to another cell type with different properties, including non-mitotic cell types. The rat ovarian granulosa cell (ROG), (Li *et al.*, 1997), and the SFME cell line (Loo *et al.*, 1987; Loo *et al.*, 1989) are examples of this type of cell. Many cancer cells also lose or gain hormone responsiveness when grown in serum, causing them to be different from the cells in the original tumor. While cell lines grown in serum have sometimes been established from these types of cells, the resulting lines do not truly represent the original cell type and are often said to be "dedifferentiated" or to have lost functionality. It is a rare event to get an immortal line from these cells, when they are cultured in serum-containing media, without transformation, which further alters the properties of the cells.

Dividing rodent (rat, mouse or hamster) cells established and maintained in serum-free medium do not undergo senescence, and do not require transformation for continuous growth in culture (Loo *et al.*, 1987).

2.2.4 Class IV. Cells That Are Killed by Components of Serum

The neuroepithelial cell line (NEP) is an example of this type of cell (Li *et al.*, 1996a). For obvious reasons, there are very few of these lines currently in existence. The only way to obtain cell lines from cell types that are killed, growth inhibited, or differentiated by serum is to place the cells in primary culture without serum, and continue with subcultur-

Table 1. Properties of NEP and ROG cells pre- and post differentiation in vitro

Pre-differentiated phenotype	Post-differentiated phenotype
ROG: rat ovarian pre-granulosa cell	
Activin required for survival	FSH required for survival
Doubling time 72 h.	Do not divide
FSH induces differentiation	Steroid response to FSH and activin
Low steriod production	1,000 x increased progesterone secretion
FSH receptor levels normal	
NEP: neuroepithelial precursor cell line	
Epithelial phenotype	Neuronal phenotype
Doubling time 48 h	Do not divide
Nestin positive	Neurofilament 200 positive
Vimentin negative	Vimentin positive
	Neuron specific enolase positive
Negative for all neuronal markers	Glutamate positive
listed at right	Aspartate positive
	Tubulin β positive
	Protein kinase C- (1.12) positive

ing, selection, etc. all in the complete absence of serum. In fact, these last two classes of cells cannot be established or grown at any point in serum supplemented medium. Each cell type will have it's own specific requirements for hormones, growth factors, substrate, nutrient concentrations, and sometimes cell shape and cell-cell contact. Very few labs work with these types of cells.

The cell lines described in Tables 1 and 2, predominantly fall into the class III & IV categories, the most rare. These types of cells are, by definition, different cell types from any of the numerous cell lines available which have been grown or initially established in serum-supplemented media. Even those cell lines that are not killed by serum benefit by being grown in serum-free conditions to promote phenotypic stability and control during passaging.

Table 2. Binding of Mabs raised against BUD and RED cells as immunogen to selected normal rat and mouse cell lines. Cell surface binding of antibodies bound to intact viable cells was measured using a FITCI-labelled second antibody and analyzed by FACS. All of the Mabs bind to the pancreatic epithelial cells from rat (BUD RED) and mouse (NODD) as well as to a neonatal rat lung epithelial cell line from respiratory bronchioles (BR516). The 5 Mabs on the left show no binding to rat cell lines from capillary endothelium (TR-1) peritubular cells (TRM) embryonic Schwann cells (ESC) or primary cultures of cardiomyocytes (rCM). The 3 Mabs on the right show some binding to the non-related cell lines although at only 1–10% of the level seen with the endoderm-derived cell lines. Each "+" represents a 10 fold increase in binding

McAb	2101	2103	2104	2160	2161	2115	2117	2140
rBUD pancreas	+++	+++	++	+++	+++	++	+	+++
rRED pancreas	+++	+++	+++	+++	+++	++	+	+++
mNODD pancreas	+++	+++	+++	++	++	++	++	++
rBR516 lung	+++	+	+++	++	+	++	+	+++
rTR-1 endothelial	–	–	–	–	–	++	+	++
rTRM peritubular	–	–	–	–	–	+	+	+
RESC Schwann	‘	–	–	–	–	+/–	+	–
rCM myocytes	–	–	–	–	–	–	–	–

2.2.5 Media Optimization

Initially a medium is optimized for the growth of each different cell type that we wish to establish from primary culture. These media were optimized for nutrient mixtures, attachment factors, hormone and growth factor supplements, and passage and subculture protocols. These will be optimized to select for the specific cell of interest and to maintain its growth rate at an optimal level for an extended number of passages. The goal is to derive conditions that will support the cell's survival and growth *in vitro* in conditions as close as possible to those that are normal for that cell *in vivo*. This means that the cells do not have to go through a prolonged adaptation or selection period to grow *in vitro*. Culture properties have been compared at the primary culture stage, early passage (p. 2 or 3) and later passage (for example p10 or 20) with very similar results from early to late passages (Li *et al.*, 1996b).

Experimental protocols may require the development of a second or third medium formulation specifically designed to support a defined cell functional activity, or differentiation, other than rapid growth. Two examples of such cells lines and their pre and post- differentiation phenotypes are summarized in Table 1. By changing the ovarian ROG cell line medium formulation, we can cause the cells to increase their steroidogenic function and decrease their replicative capacity (Li *et al.*, 1997). Another example is the differentiation of the neuroepithelial precursor cell line, NEP, to mature non-mitotic cells with the phenotype of CNS neurons by the addition of the appropriate growth/differentiation factors (Li *et al.*, 1996a).

2.3 Cell Surface Proteins: Important Regulators of Survival, Growth and Development

In multicellular organisms, the need to coordinate the activities of one cell with those of its neighbors has resulted in the evolution of complex cellular interactions that involve secreted polypeptides, cell surface proteins and extracellular matrix components. Normal epithelial cell surfaces present a diverse and multifunctional array of membrane proteins that mediate hormone-cell, cell-cell and cell-matrix interactions (Edelman, 1986; Takeichi, 1988). Cellular activities are modulated in response to signals received from neighboring cells and the immediate environment (van der Geer *et al.*, 1994). Among the cellular interactions, cell-cell and cell-matrix interactions represent complex and dynamic forms of communication which provide information to the cells for controlling morphogenesis, cell fate specification, gain or loss of tissue-specific functions, cell migration, cell proliferation, tissue regeneration, and cell death (Werb, 1997).

Most cloning efforts to date have dealt with secreted extracellular signaling molecules such as hormones, growth factors, and cytokines. However, more recently, proteins which are either wholly or partially membrane bound, such as the neuregulins (Zhang *et al.*, 1998), patched, and hedgehog families of proteins (Echelard *et al.*, 1993; Johnson *et al.*, 1996), have been recognized as playing a crucial role in regulating development. An approach biased towards discovering cell surface proteins present during the development of an organ would, therefore, provide new information on the regulation of that development.

2.4 Raising Antibodies to Cell Surface Proteins. A New Approach to an Established Technology Yields Unique Set of Antibodies

Several approaches have been used to identify cell surface proteins. Signal sequence trapping, subtraction cloning, differential display, SAGE, and genechip technology have become popular for protein discovery and cloning. Hybridoma technology remains an effective and a rapid process for identification and validation of cell type-, developmental stage-, and disease-specific antigens. This approach can be designed to be particularly useful for identifying, cloning, and studying cell surface proteins.

2.4.1 Why Use Mabs As a Primary Approach to Identifying CSAs?

A biologically active Mab allows some understanding of the possible biological role of the antigen before the expenditure of much time or effort on purifying or cloning a protein of unknown function. In addition to helping to characterize the biological role of the antigen, the monoclonal antibody, in itself, provides a path to rapid expression cloning of the antigen it recognizes, and /or allows purification of the antigen using affinity column chromatography.

One can clone the genes coding for the antigen recognized by the Mab by making a library from the cell line used to raise the antibody, and expressing this library in COS cells or another host line that does not contain the antigen. In many cases, the gene coding for the antigen can then be rapidly isolated and sequenced using a modification of the panning method of Aruffo and Seed (Aruffo and Seed, 1987; Stephan *et al.*, 1999a). This method is ideal for cell surface proteins and antibodies that recognize the extracellular region of such proteins. In some cases, for example, if the protein is coded by non-contiguous gene sequences, the panning method will not yield a clone. In these cases affinity purification of the protein followed by sequencing the protein and searching the protein and DNA sequence databases for homology, will facilitate cloning.

Finally a human, or humanized, anti-human Mab, which is well tolerated and has a long half-life *in vivo*, itself has great potential as a therapeutic product. Monoclonal antibodies have currently been approved for the treatment of cancer, prevention of blood clots after angioplasty, preventing respiratory disease in newborns, treating Crohn's disease and other helping prevent organ rejection after transplant surgery.

2.4.2 Why Progenitor Cells Provide a Significant and Unique Source of Antigens

Immunizing mice with embryonic cells to prepare monoclonal antibodies is not a new concept. However, the concept can be efficiently reduced to practice *only* when the technology exists which will allow embryonic cells to expand as a homogeneous population in reasonably large scale without differentiation or transformation. Our human embryonic cell lines are derived from the liver, ovary, pancreas, endometrium, brain, and lung. Defined culture conditions allow the selection and derivation of cell lines "arrested" in a specific stage of development. These homogeneous cell lines exhibit developmental stage-specific arrest and maintain characteristics (including antigen expression) of the developmental stage from which the cell line was initially established. We employ normal embryonic undifferentiated cell lines representing cell types not previously cultured in any laboratory, to provide sources of novel antigens which might not be expressed at high levels in normal adult tissue. These cell lines are an unlimited source of consistent, pure, fetal-antigen-expressing, cells for immunization, screening, and *in vitro* bioassays.

Many of the genes important in regulating development (or cellular regeneration) are normally expressed at high levels only early in differentiation, and would not be found in normal adult cells.

From this point of view, stem cells, progenitor cells, or precursor cells will provide rich sources of novel antigens, many of which may play crucial roles in regulating cell growth, survival, and differentiation.

2.4.3 Advantages of Using Intact Cells As Antigen

Using intact cells as the antigen provides two advantages compared to using purified proteins or tissue extracts. The cell surface proteins (eg a receptor) are presented in a biologically relevant manner. The three-dimensional configuration of the protein is intact and the active portion of the molecule is exposed, while most of the molecule is buried in the lipid bi-layer or is inside the cell. To date, all Mabs tested raised using our methodology have been biologically active, i.e. they can block a biological event such as cell growth or differentiation (Stephan *et al.*, 1999a; Stephan *et al.*, 1999b).

Having a cell line arrested at a unique stage of development provides a well defined, reproducible source of cell surface antigens that may be rare or absent in adult tissues. Since the lines represent only one type of cell, the antigen of interest may be highly expressed in these cells, but not well represented in a complete tissue. Using cells as antigens has long been considered as a way to generate antibodies to cell surface proteins. However, this method has generally proven less useful than expected, due to the high level of cross-reactivity of the antibodies obtained with most other cultured cells. This was, at least in part, due to the serum proteins bound to the cell surface. Our method of raising antibodies has yielded no universally cross-reactive antibodies to date.

2.5 An Example: Immunization with Intact Rat Fetal Pancreatic Cells

During the last few years, the epithelial cell layer of the developing pancreatic bud gives rise to all of the cells of the endocrine pancreas (islets) and the exocrine (ducts and acini) pancreas. In addition, 80% of all pancreatic cancer arises from the ductal epithelium. Cell lines derived from these cells would therefore be very interesting tools for understanding the development and regulation of the function of the pancreas during normal development and in pathological events.

2.5.1 Establishing Cell Lines from the Fetal Pancreas

Two pancreatic epithelial cell lines were established from primary cultures of dissected rat e12 embryonic pancreatic buds (BUD) and rat e17 ductal epithelium (RED), respectively. BUD cultures were established by surgically dissecting the dorsal and ventral pancreatic evaginations and culturing each in separate wells of a 48 well dish without initial enzymatic dissociation of the tissue. The cells were plated on fibronectin-coated plates in growth medium which consisted of F12/DME supplemented with 14F: rhu-insulin, transferrin, EGF, ethanolamine, aprotinin, additional glucose, phosphoethanolamine, triiodothyronine, selenium, hydrocortisone, progesterone, forskolin, heregulin $\beta_{177-244}$, and bovine pituitary extract (BPE). Pancreati from older embryos were enzymatically digested and the ductal pieces partly purified and cultured as described above. The cultures are initiated and carried in a serum-free 14F medium optimized to select for the growth of the epithelial cells. Each component of the medium contributes to the optimal growth of the cells. Under these optimal conditions, the fibroblast and mesenchymal cells are lost from the cultures within 2 passages and the remaining cells are uniformly epithelial. The cultures have a log phase population doubling time of 11.4 hr and 14 hr for BUD and RED cells, respectively (Stephan *et al.*, 1999b). The cells form a contact-inhibited monolayer, have a normal karyotype and have been grown continuously for over 80 population doublings with no obvious change in cell morphology or growth profile. In accordance with previous work establishing rodent cell lines in this fashion (Loo *et al.*, 1989; Roberts *et al.*, 1990), no cell senescence has been observed.

In order to better characterize the BUD and RED cell lines, the presence of various proteins known to be present at early stages of pancreatic development were investigated by Western blot analysis. We demonstrate that the BUD cells and, to a lesser extent, the RED cells express cytokeratin 7, which is present only in the pancreatic ductal epithelium (Bouwens, 1998). BUD and RED cells also express carboxypeptidase A, another ductal marker (Kim *et al.*, 1997). The procarboxypeptidase is present in the two pancreatic cell lines. Both pancreatic BUD and RED cells also express the mRNA coding for homeodomain-containing transcription factor for insulin gene expression PDX1, which appears in the pancreatic bud epithelium early, before insulin, in the ontogeny of the pancreas (Watada *et al.*, 1996). In the adult, PDX1 is expressed only in β-cells and not in the mature ductal epithelium. Tyrosine hydroxylase, a marker for early islet progenitor (Teitelman *et al.*, 1987) and early ductal cells, was detected in the BUD cells and to a lesser extend in the RED cells.

2.5.2 Raising Antibodies to Cell Surface Antigens

The BUD and RED cells lines were used as immunogens to raise a panel of antibodies that are specific for the native configuration of cell surface molecules present on these fetal cells. Fifteen Mabs, recognizing 13 distinct antigens were obtained. The binding properties of eight of these to normal rat and mouse cell lines are shown in Table 2.

Each of these antibodies binds to the developing pancreatic epithelium *in vivo*, some exclusively so (Stephan *et al.*, 1999b). This data confirms that, in addition to the known cytoplasmic markers, the BUD and RED cells have maintained the presentation of a set of cell surface antigens appropriate to the cell type and stage of development of the cells from which they were derived. These data show that we can use the cell surface proteins recognized by these and other Mabs as an additional set of markers for pancreatic development. Cloning of the genes coding for the antigens recognized by three of the antibodies allowed us to identify them as the rat homologue of EpCAM (2160), rat BEN (2117), and the rat homologue of the alpha 1,2-fucosyltransferase enzyme (2103). Two of these had not previously been identified as markers for the developing pancreas (Stephan *et al.*, 1999a; Stephan *et al.*, 1999b).

2.6 Novel Cell Lines and Panels of Mabs to Cell Surface Proteins Provide New Research and Drug Discovery Tools

The above example shows the results using two rat cell lines as immunogens. We subsequently derived a human fetal pancreatic epithelial cell line and used this as an immunogen. As with the rat cell lines, a panel of Mabs recognizing several distinct cell surface antigens present on fetal pancreatic epithelial cells was obtained. We have raised panels of antibodies to both rat and human, normal and tumorigenic, embryonic and adult cell lines. Based on data generated to date, we can generalize that the Mabs that we obtain using our protocol have the following unique set of properties:

- All recognize proteins present on the surface of the cell
- All recognize the native configuration of these CSAs
- All tested so far have been able to block a biological activity such as cell division or differentiation of cultured organs *in vitro* (Stephan *et al.*, 1999a; Stephan *et al.*, 1999b)
- None of the Mabs recognize universally expressed CSAs, but are related to the antigenic cell type used
- Mabs have been raised to cell surface receptors, attachment proteins and enzymes, suggesting that many classes of active proteins can be targeted using this approach

Because the Mabs obtained are biologically relevant, they have multiple uses in drug development and target discovery including use as: 1) therapeutic human monoclonal antibodies; 2) tools for the identification and cloning of the genes coding for cell surface antigens, allowing subsequent production of recombinant soluble proteins containing the extracellular domains for cell surface antigens for further study; and 3) tools for diagnostics and *in vivo* imaging. They will also provide valuable tools for basic research into cellular regulation, signaling, and development.

These cell lines represent pure cell types at specific stages of differentiation. Many of them can be synchronously differentiated *in vitro* in a defined manner to a later, non-mitotic stage of development. They therefore provide excellent model systems for applying other recently developed technologies to study the process of differentiation in a controlled system. Many of the cell lines secrete a large number of proteins. The Schwann cell lines (rESC and rASC) secrete over 300 spots on a 2-dimensional gel (unpublished data). These cell lines might be used as the starting point for proteomics technologies of microsequencing and identification of proteins using mass spectroscopy to identify and analyze protein secretion in response to drugs or physiologically relevant hormones and growth factors (Arnott *et al.*, 1998). In parallel, the cultures provide an excellent source of material for RNA profiling or differential gene cloning as the cells differentiate and/or respond to specific hormone signals *in vitro*.

It has been known for many years that many cancers "overexpress" or "turn on" fetal genes. These fetal tissue stem cell lines might then become a rich source of discovering targets for therapeutic monoclonal antibodies to treat cancers. Finally there is a growing body of evidence that embryonic remnant cells in the adult may lead to diseases ranging from some cancers, for example, teratocarcinomas to endometriosis (Mai *et al.*, 1998). Having cell lines with the characteristics of these fetal populations would help understand the defect leading to the persistence of these disease-causing cells in the adult.

References

Arnott D, O'Connell K, King K, Stults J (1998) An integrated approach to proteome analysis: identification of proteins associated with cardiac hypertrophy. Anal Biochem 258: 1–18

Aruffo A, Seed B (1987) Molecular cloning of a CD28 cDNA by a high-efficiency COS cell espression system. Proceeding of the National Academy of Sciences. USA 84: 8573–8577

Barnes D, Sato G (1980) Serum-free cell culture: a unifying approach. Cell 22: 649–55

Bottenstein J, Hayashi I, Hutchings SH, Masui H, Mather J, McClure DB, Okasa S, Rizzino A, Sato G, Serroro G, Wolfe R, Wu R (1979) "The growth of cells in serum free hormone supplemented media." Academic Press, New York

Bouwens L (1998) Cytokeratins and cell differentiation in the pancreas. Journal of Pathology 184: 234–239

Echelard Y, Epstein DJ, St-Jacques B, Shen L, Mohler J, McMahon JA, McMahon AP (1993) Sonic hedgehog, a member of a family of putative signaling molecules, is implicated in the regulation of CNS polarity. Cell 75: 1417–1430

Edelman GM (1986) Cell adhesion molecules in the regulation of the animal form and tissue pattern. Annual Review of Cell Biology 2: 81–116

Frederiksen K, Jat PS, Valtz N, Levy D, McKay R (1988) Immortalization of precursor cells from the mammalian CNS. Neuron 1: 439–448

Johnson RL, Rothman AL, Xie J, Goodrich LV, Bare JW, Bonifas JM, Quinn AG, Myers RM, Cox DR, Epstein EH, Scott MP (1996) Human homolog of patched, a candidate gene for the basal cell nevus syndrome. Science 272: 1668–1671

Kim SK, Hebrok M, Melton DA (1997) Notochord to endoderm signaling is required for pancreas development. Development 124: 4243–4252

Li R.-H, Phillips D, Moore A, Mather J (1997) Follicle-stimulating hormone induces terminal differentiation in a pre-differentiated rat granulosa cell line (ROG) Endocrinology 138: 2648–2657

Li RH, Gao WQ, Mather JP (1996a) Multiple factors control the proliferation and differentiation of rat early embryonic (day 9) neuroepithelial cells. Endocrine 15: 205–217

Li RH, Sliwkowski MX, Lo J, Mather JP (1996b) Establishment of Schwann cell lines from normal adult and embryonic rat dorsal root ganglia. Journal of Neuroscience Methods 67: 57–69

Loo D, Fuquay J, Rawson C, Barnes D (1987) Extended Culture of Mouse Embryo Cells Without Senescence: Inhibition by Serum. Science 236: 200–202

Loo D, Rawson C, Helmrich A, Barnes D (1989) Serum-free mouse embryo cells: growth responses in vitro. Journal of Cellular Physiology 139: 484–491

Mai K, Yadzi H, Perkins D, Parks W (1998) Development of endometriosis from embryonic duct remnants. Hum Pathol 29: 319–322

Mather J, Barnes D (1998) Animal Cell Culture Methods. In: Wilson L, Matsudaira P (eds) Methods in cell biology, vol 57. Academic Press, New York

Mather J, Roberts P (1998) "Introduction to Cell and Tissue Culture: Theory and Technique." Plenum Press, New York

Morrissey T, Levi A, Niujens A, Sliwkowski M, Bunge R (1995) Axon-induced mitogenesis of human Schwann cells involves heregulin and p185erbB2. Proc Natl Acad Sci USA 92: 1431–1435

Murakami H, Yamane I, Barnes D, Mather J, Hayashi I, Sato G (1985) Growth and Differentiation of Cells in Defined Environment. Springer, New York

Roberts PE, Phillips DM, Mather JM (1990) Properties of a novel epithelial cell from immature rat lung: Establishment and maintenance of the differentiated phenotype. Amer J Physiol: Lung Cell & Molec Physiol 3: 415–425

Stephan JP, Lee J, Bald L, Gu Q, Mather JP (1999a) Distribution and function of the adhesion molecule BEN during development. Developmental Biology 212: 264–277

Stephan JP, Roberts PE, Bald L, Lee J, Gu Q, Devaux B, Mather JP (1999b) Selective cloning of cell surface proteins involved in organ development: EGP is involved in normal epithelial differentiation. Endocrinology 140: 5841–5854

Takeichi M (1988) The cadherins: cell-cell adhesion molecules controlling animal morphogenesis. Development 102: 639–655

Teitelman G, Lee JK, Alpert S (1987) Expression of cell type-specific markers during pancreatic development in the mouse: implications for pancreatic cell lineages. Cell Tissue Res 250: 435–9

Thomson JA, Itskovitz-Eldor J, Shapiro SS, Waknitz MA, Swiergiel JJ, Marshall VS, Jones JM (1998) Embryonic stem cell lines derived from human blastocysts. Science 282: 1145–7

van der Geer P, Hunter T, Lindberg RA (1994) Receptor protein-tyrosine kinases and their signal transduction pathways. Annual Review of Cell Biology 10: 251–337

Watada H, Kajimoto Y, Miyagawa J, Hanafusa T, Hamaguchi K, Matsuoka T, Yamamoto K, Matsuzawa Y, Kawamori R Yamasaki Y (1996) PDX-1 induces insulin and glucokinase gene expressions in aTC1 clone 6 cells in the presence of betacellulin. Diabetes 45: 1826–1831

Werb Z (1997) ECM and cell surface proteolysis: regulating cellular ecology. Cell 91: 439–442

Zhang D, Frantz G Godowski PJ (1998) New branches on the neuregulin family tree. Molecular psychiatry 3: 112–115

3 The Wonderful World of Proteomics

K. L. Bennett, E. Mørtz

3.1 Introduction 27
3.2 Classical or Expression Proteomics 28
3.3 Functional Proteomics 30
3.4 Proteomics and Mass Spectrometry 30
3.5 Perspectives 37
References 37

3.1 Introduction

3.1.1 Definition

The term *proteome* is defined as: *The entire protein complement expressed by a genome, a cell or tissue type* (Wasinger et al. 1995, Wilkins et al. 1996). It follows that *proteomics* is the study of proteomes, however the terminology now encompasses a broader field of research and has branched into two specific disciplines; classical and functional proteomics.

3.1.2 MDS Protana A/S

MDS Protana A/S plays a key role in the rapidly expanding world of classical and functional proteomics. The following chapter outlines the approaches we have successfully established in our laboratory (summa-

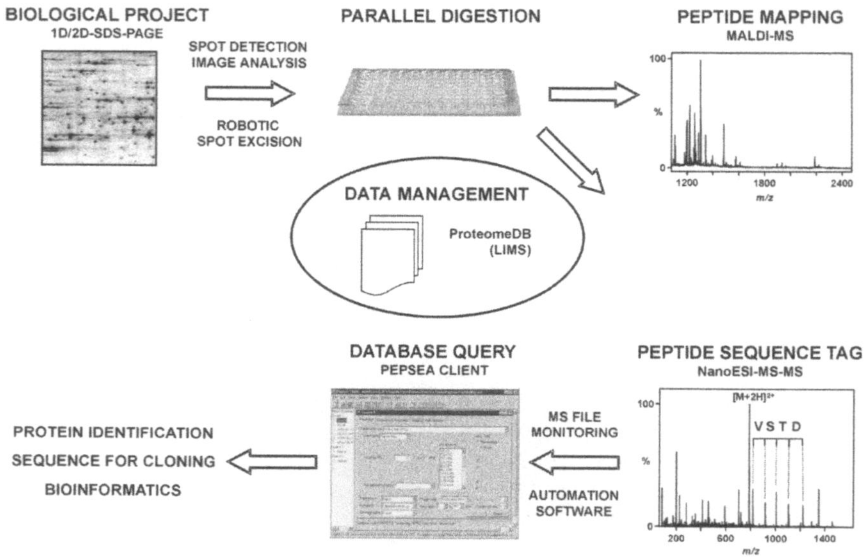

Fig. 1. General overview of the proteome approach employed at MDS ProtanaA/S. Proteins separated by 1D or 2D SDS–PAGE are excised and digested *in situ* in 96–well plates. The digested samples are analysed by MALDI–MS (*Peptide Mapping*) and/or nanoESI–MS–MS (*Peptide Sequence Tag*). The information generated by mass spectrometry automatically queries nonredundant sequence database (nrdb), EST or genome databases. Proteins are either unequivocally identified, or if novel, sufficient *de novo* sequence is generated for cloning and cDNA sequencing. The entire process is monitored by an integrated database management system (*ProteomeDB*)

rised in Fig. 1) for the evaluation of intact or partial proteomes and identification of novel proteins.

3.2 Classical or Expression Proteomics

The total protein complement of two or more differentially–treated cell or tissue lines are separated on the basis of isoelectric point (pI) and molecular mass by two dimensional sodium dodecyl sulphate polyacrylamide gel electrophoresis (2D–SDS–PAGE). The gels are compared and differences in protein expression between the reference and altered states are determined by the up– or down–regulation of particular proteins. Large–scale 2D gel electrophoresis is combined with computer–based image analyses to quantify the ratio of spot intensities and thereby determine global changes between the proteomes. Once differentially–displayed proteins are detected, the proteins are identified by mass spectrometry (MS). Little to no prior knowledge concerning the proteome under investigation is required, however, there are several disadvantages of classical proteomics. 2D gels only permit the visualisation of approximately 20 % of the loaded protein; only proteins of molecular mass 10 kDa to 100 kDa migrate satisfactorily within 2D gels; membrane proteins are often poorly analysed; relevant differences in protein patterns between differentially–treated proteomes are difficult to establish due to the variability of 2D gels; apparent expression differences may be irrelevant to the actual experiment; and proteins present in low–copy–number (10–1000 copies per cell) are not detected. In addition to these inherent problems, spot recognition and quantification is often time–consuming and the accuracy of the technique is limited. In spite of the disadvantages outlined above, classical proteomics still provides an invaluable means of viewing the majority of the proteins in a cell extract or tissue and is particularly beneficial for mapping previously–uncharacterised cell lines.

Classical proteomics, however, does not permit the specific function of a protein to be determined. A complete and thorough understanding of function is mandatory to comprehend biochemical pathways or to design and develop new drugs. At present these processes are hampered by the inability to systematically determine the role of all proteins within a cell. Proteins rarely act independently but are more often found as components of multiprotein complexes which cooperate to drive cellular machinery and perform specific roles within the cell. Information on a protein can often be inferred from associated *protein partners* or from its subcellular location, therefore, systematic identification of individual components of protein complexes should permit definition of these *cellular machines*. From the generated information, it is feasible that *maps* of different cell types and states can be constructed. Classical or expression proteomics does not provide all the necessary requirements to address such issues and thus obviates the need for the second of the proteome approaches, i.e., functional proteomics.

3.3 Functional Proteomics

An important criterion for successful functional proteomics, is that a basic knowledge of at least one of the proteins under investigation is essential. Once this is established, or at least postulated, systematic *gene tagging* with peptide sequences (e.g., by fusing to GST) or immunoprecipitation with specific antibodies enables affinity isolation of a subset of proteins from a cell or tissue extract. As nondenaturing conditions are employed, even weakly– or transiently–associated proteins are successfully sequestered. Each protein in the purified subset has a common feature which is exploited in the isolation procedure, e.g., the proteins share an affinity for a common target. Following isolation, the protein subset is visualised by single (1D) or two dimensional (2D) sodium dodecyl sulphate polyacrylamide gel electrophoresis and the separated proteins identified by mass spectrometry (MS). After bioinformatic validation, the trait common to all the proteins in the complex provides information on the function of the identified proteins. By improving our understanding of protein–protein interactions, novel proteins are placed in their functional context, e.g., within a biochemical pathway; as components of a multiprotein complex; or as constituents of subcellular machinery. Functional proteomics deals with simplified systems, therefore many of the disadvantages of the classical proteome approach are alleviated. Monitoring a reduced subset of proteins (as opposed to intact proteomes) is markedly simplified and differences in protein expression levels are easier to detect in differential display experiments. In addition, the practical sensitivity limit is high (low femtomolar quantities isolated); and the detection of low copy number proteins is possible as a consequence of affinity purification enrichment.

3.4 Proteomics and Mass Spectrometry

The expeditious advancement of the world of proteomics is largely due to the emergence of biological mass spectrometry (MS). MS has revolutionised the analysis of intact or subset proteomes by providing rapid and sensitive identification of proteins in 1D and 2D gels. The major advantages of identifying proteins via MS compared to traditional methods include: high sensitivity (femtomolar levels) thereby allowing analysis of Coomassie Brilliant Blue – and silver–stained proteins; high specificity which provides unique protein identification; mass determination of several peptides from a protein resulting in verification of significant region(s) of the protein sequence; identification of several proteins in a mixture; and parallel processing of large numbers of samples.

3.4.1 MALDI–MS Peptide Mapping and Database Searching

To identify proteins from 1D and 2D gels, protein spots are excised and placed in 96–well plates (Fig. 1). The gel plugs are washed and cysteine residues reduced and alkylated. The proteins are cleaved *in situ* with trypsin (or another specific endoproteinase) and the generated peptides (generally in the mass range of 500–2500 Da) are eluted from the gel pieces. An aliquot (2 %) of each sample is analysed on a Bruker REFLEX III MALDI reflector–time–of–flight (MALDI–RE–TOF) mass spectrometer (Bruker–Daltronics, Bremen, Germany) equipped with a SCOUT 384 multiprobe inlet and a gridless delayed–extraction ion source. The spectra are acquired in automatic mode using a *fuzzy logic control algorithm* to regulate the laser fluence (Jensen et al. 1997), thereby enabling rapid analysis of large numbers of samples (i.e., high–throughput protein characterisation). The combination of pulsed ion extraction MALDI–MS with a reflecting TOF analyser has resulted in mass accuracies in the low parts per million (30 ppm) (Jensen et al. 1996) and this feature is highly advantageous for accurate mass determination. The experimentally–determined mass spectrometric *protein fingerprint* or *peptide map* (Fig. 1) generated from *in situ* digestion of the protein is queried against a nonredundant sequence database (nrdb) containing all known protein sequences (approximately 430,000 to date). The measured peptide masses are compared to a theoretical proteolytic digest of all the proteins in the database. A score representing the similarity between the experimental and the theoretical peptide masses i.e., the number of matching peptide masses and the mass accuracy used in the search, is calculated for each protein. The protein is identified as the one with the closest agreement between the measured and calculated peptide mass maps. In principle, the protein with the highest score should result in the correct identification.

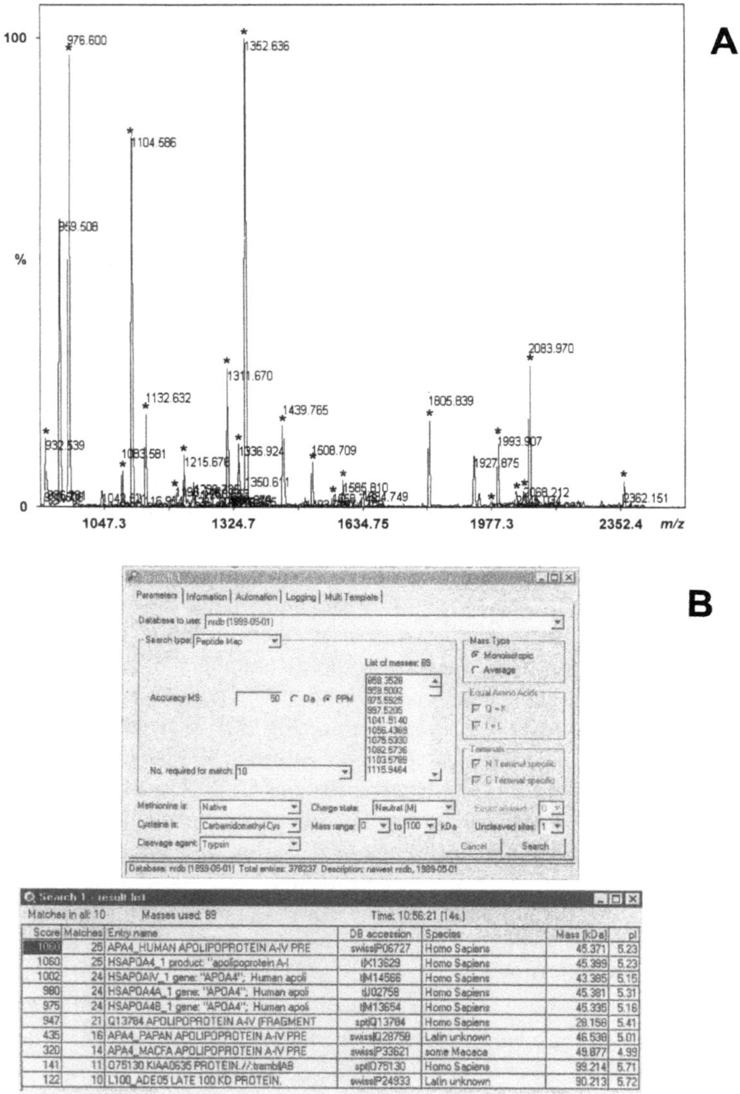

Fig. 2 A,B. Legend see p. 33

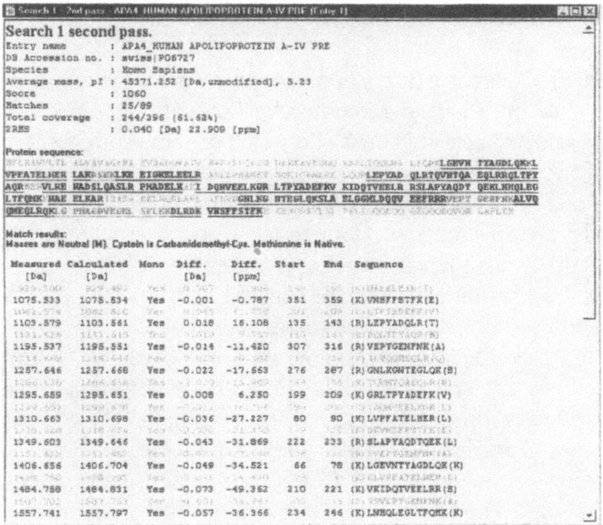

C

Fig. 2A–C. Identification of an unknown protein by MALDI-MS peptide mapping. (A) MALDI–TOF peptide map obtained from an *in situ* tryptic digest of the unknown protein; (B) peptide masses determined from the MALDI mass spectrum are queried against a nonredundant sequence database (nrdb) using the PepSea Client software and the protein was unequivocally identified as human apolipoprotein; (C) confirmation of the identification was achieved by correlating the matching peptide masses with the protein sequence in a *Second Pass Search*

Shown in Fig. 2 A is the MALDI mass spectrum obtained from an *in situ* tryptic digest of an unknown protein. The peptide masses were queried against the nonredundant sequence database using PepSea Client Version 2.2.3 (MDS Protana A/S, Odense, Denmark) and the unknown protein was unequivocally identified as human apolipoprotein (Fig. 2B). Identified proteins are confirmed by correlating the matching peptide masses with the protein sequence by a *Second Pass Search* (Fig. 2 C). Not all samples, however, are identified by the peptide masses determined by MALDI–TOF–MS. This situation can arise if the number of trypsin–generated peptides is too low for the database query; if the gel piece contains more than a single protein; or the protein is novel and therefore the sequence is not present in the database.

3.4.2 Peptide Sequence Tags and Database Searching

A more specific database query can be performed by using partial peptide sequence information generated by tandem mass spectrometry (MS–MS) of selected peptide peaks (Fig. 1). The remaining 98 % of the sample from the *in situ* tryptic digest is desalted and concentrated on a disposable microcolumn (glass capillary or GELoader tip, Eppendorf–Netheler–Hinz GmbH, Hamburg, Germany) packed with POROS R2 20 reverse–phase media (PerSeptive Biosystems, Cambridge, MA). The bound peptides are washed with 10–20 µL 5 % formic acid and eluted into precoated borosilicate nanoelectrospray needles (MDS Protana A/S, Odense, Denmark) with 0.3–1 µL 50 % aqueous methanol, 5 % formic acid. Peptide sequences are determined using a quadrupole reflector–time–of–flight mass spectrometer (Q–STAR) (PE Sciex, Toronto, Canada) equipped with a nanoelectrospray source (MDS Protana A/S, Odense, Denmark). The combination of nanoESI–MS with a quadrupole analyser followed by a collision cell and a reflecting TOF analyser (Morris et al. 1996, Shevchenko et al. 1997) not only permits higher sensitivity and resolution but also enables sample volumes as low as 300 nL with flow rates ranging from 10–40 nL/minute.

Illustrated in Fig. 3 is the nanoESI mass spectrum of an *in situ* tryptic digest of 100 fmoles bovine serum albumin (BSA) (Fig. 3 A) and the nanoESI–MS–MS spectrum of the doubly–charged peptide at *m/z* 653.38 (Fig. 3B). Peptides in the mixture are labelled **P** and **T** (BSA and trypsin autoproteolysis products, respectively). The high resolution (8,000–10,000) of the Q–STAR permits charge state determination of multiply–charged peptides and enables peptide signals to be easily discriminated from background chemical and electronic noise. Selected peptides are fragmented by tandem mass spectrometry (MS–MS), and partial amino acid sequences are deduced from the fragment ion series generated from the precursor ion (Fig. 3B). The high mass accuracy and resolution of the Q–STAR allow confident sequence assignments.

To query the nonredundant database (nrdb), the mass of the intact peptide, the partial amino acid sequence derived from the MS–MS spectrum, and the masses defining the position of the sequence within the peptide are combined into a *Peptide Sequence Tag* (Mann and Wilm

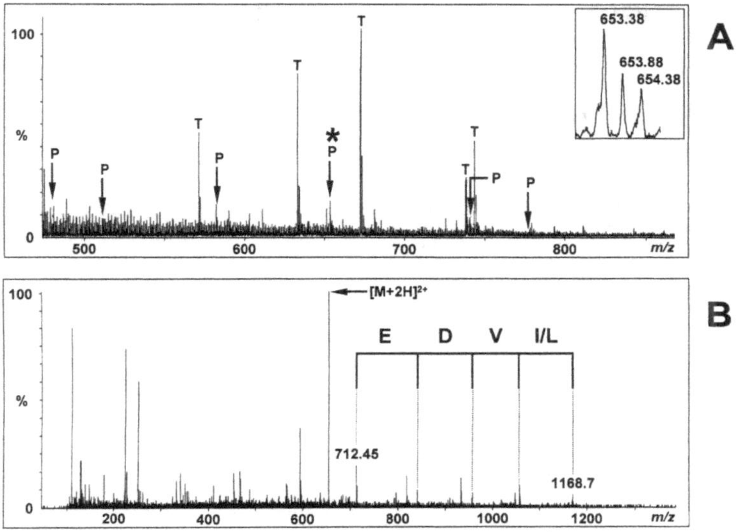

Fig. 3. *In situ* tryptic digest of 100 fmoles bovine serum albumin (BSA). (A) NanoESI mass spectrum of the BSA digest. Peptides in the mixture are labelled **P** and **T**, which correspond to BSA and trypsin autoproteolysis products, respectively. Shown inset is the BSA peptide marked with an asterix. (B) NanoESI tandem mass spectrum of the doubly–charged ion at *m/z* 653.38. The partial amino acid sequence combined with the mass of the intact peptide and the masses defining the position of the sequence within the peptide (*Peptide Sequence Tag*) provide sufficient information to identify the protein

1994). Unique hits in the database are often obtained with a partial sequence comprised of 3–4 amino acid residues. When a peptide is identified, the theoretical fragmentation is predicted and compared to the MS–MS spectrum for assignment of peaks that can validate the identification. This procedure is repeated for all fragmented peptides and provides either additional verification of the result, or identification of additional proteins.

3.4.3 EST and Genome Database Searching – Novel Protein Identification

The identification of a protein requires that the sequence is available in the nonredundant database. For the majority of organisms, many proteins are likely to be unknown. Novel proteins are of particular interest for target discovery and validation, or as candidates for *missing links* in biochemical pathways. The initial step in ascertaining the biological function of a novel protein is to generate sufficient sequence information for cloning and cDNA sequencing. Generation of *de novo* sequence information is achieved with the use of the Q–STAR mass spectrometer (PE Sciex, Toronto, Canada) equipped with nanoelectrospray source. Multiple sequence stretches of up to 15–20 amino acid residues are produced by nanoESI–MS, which is more than sufficient information for the subsequent construction of oligonucleotide probes. The mass spectrometric sequence is used to search public expressed–sequence tag (EST) and genome databases. If one or several nucleotide sequences are identified, long sequences will be provided which can be used for primer construction. The *Peptide Sequence Tag* algorithm has been adapted to allow direct screening of EST (Neubauer et al. 1998) and genome databases (Küster et al. submitted). Translating the amino acid sequence used to construct the peptide sequence tag into the corresponding degenerate oligonucleotide sequence avoids the need to translate the EST databases into different reading frames. EST clones are confirmed by matching multiple peptide sequence tags either to the initial database sequence, or to extended sequences obtained by assembling independent ESTs and by direct sequencing of EST clones. An increasing number of novel proteins are being identified by this approach. The most prominent are FLICE, the elusive signalling molecule between the apoptosis receptor Apo1/Fas and the ICE protease family (Muzio et al. 1996); IKK–1 and IKK–2, the cytokine–activated kinases essential for activation of NF–κB (Mercurio et al. 1997); and the components of the multiprotein spliceosome complex (Neubauer et al. 1998).

3.5 Perspectives

The successful amalgamation of proteomics with mass spectrometry, accompanied by expanding EST and genome databases and improved search algorithms, will ensure that the characterisation of novel proteins and determination of protein function will continue. The subcellular location of proteins and verification of specific protein–protein interactions will permit the construction of a three–dimensional *virtual cell* (Blackstock and Weir 1999). This will not only supply cell biologists with an unprecedented wealth of information concerning the protein composition of any isolated cellular or macromolecular structure, but it will also provide the basis for discovery of target molecules in the search for new drug design and for the elucidation of new biochemical pathways.

References

Blackstock WP, Weir MP (1999) Proteomics: quantitative and physical mapping of cellular proteins. Trends Biotechnol 17:121–127

Jensen ON, Podtelejnikov A, Mann M (1996) Delayed extraction improves specificity in database searches by matrix-assisted laser desorption/ionisation peptide maps. Rapid Commun Mass Spectrom 10:1371–1378

Jensen ON, Mortensen P, Vorm O, Mann M (1997) Automation of matrix-assisted laser desorption/ionisation mass spectrometry using fuzzy logic feedback control. Anal Chem 69:1706–1714

Kuster B, Mortensen P, Mann M (2000) Mass spectrometry allows direct identification of proteins in large genomes. Science Submitted

Mann M, Wilm M (1994) Error-tolerant identification of peptides in sequence databases by peptide sequence tags. Anal Chem 66:4390–4399

Mercurio F, Zhu H, Murray BW, Shevchenko A, Bennett BL, Li J, Young DB, Barbosa M, Mann M, Manning A, Rao A (1997) IKK-1 and IKK-2: cytokine-activated IκB kinases essential for NF-κB activation. Science 278:860–866

Morris HR, Paxton T, Dell A, Langhorne J, Berg M, Bordoli RS, Hoyes J, Bateman RH (1996) High-sensitivity collisionally-activated decomposition tandem mass spectrometry on a novel quadrupole/orthogonal-acceleration time-of-flight mass spectrometer. Rapid Commun Mass Spectrom 10:889–896

Muzio M, Chinnaiyan AM, Kischkel FC, O'Rourke K, Shevchenko A, Ni J, Scaffidi C, Bretz JD, Zhang M, Gentz R, Mann M, Krammer PH, Peter ME, Dixit VM (1996) FLICE, a novel FADD-homologous ICE/CED-3-like protease, is recruited to the CD95 (Fas/APO-1) death-inducing signalling complex. Cell 85:817–827

Neubauer G, King A, Rappsilber J, Calvio C, Watson M, Ajuh P, Sleeman J, Lamond A, Mann M (1998) Mass spectrometry and EST-database searching allows characterisation of the multi-protein spliceosome complex. Nat Genet 20:46–50

Shevchenko A, Chernushevich I, Ens W, Standing KG, Thomson B, Wilm M, Mann M (1997) Rapid '*de novo*' peptide sequencing by a combination of nanoelectrospray, isotopic labelling and a quadrupole/time-of-flight mass spectrometer. Rapid Commun Mass Spectrom 11:1015–1024

Wasinger VC, Cordwell SJ, Cerpa-Poljak A, Yan JX, Gooley AA, Wilkins MR, Duncan MW, Harris R, Williams KL, Humphery-Smith I (1995) Progress with gene-product mapping of the Mollicutes: *Mycoplasma genitalium*. Electrophoresis 16:1090–1094

Wilkins MR, Pasquali C, Appel RD, Ou K, Golaz O, Sanchez JC, Yan JX, Gooley AA, Hughes G, Humphery-Smith I, Williams KL, Hochstrasser DF (1996) From proteins to proteomes: large scale protein identification by two-dimensional electrophoresis and amino acid analysis. Bio/Technology 14:61–65

4 Cyclin-Dependent Kinases and Their Chemical Inhibitors: New Targets and Tools for the Study of the Testis?

L. Meijer, M. Leost, S. Leclerc, M. Garnier

4.1 Cyclin-Dependent Kinases 39
4.2 ATP Site-Directed CDK Inhibitors 44
4.3 CDKs and Cyclins in the Testis 50
4.4 Conclusions 52
References 53

4.1 Cyclin-Dependent Kinases

Among the estimated 1,000 to 2,000 human protein kinases, a family of kinases activated by a family of cyclins, the cyclin-dependent kinases (CDKs), has been extensively studied because of their essential role in the regulation of cell proliferation, of neuronal and thymus functions and of transcription (reviews in ref. [1–4]). The first identified CDK, cdc2, was initially discovered as a gene essential for both G1/S and G2/M transitions in yeast [5]. Following the cloning of the human cdc2 homologue, CDK1, by complementation [6], cdc2 homologues were found to be present in all eucaryotes from plants and unicellular organisms to humans. It was also realised that cdc2 was only the first member of a family of closely related kinases (Fig. 1). Following the initial discovery of cyclin B in sea urchin eggs, it was also shown that cyclin B homologues were present in all eucaryotes, and that, here again, it was

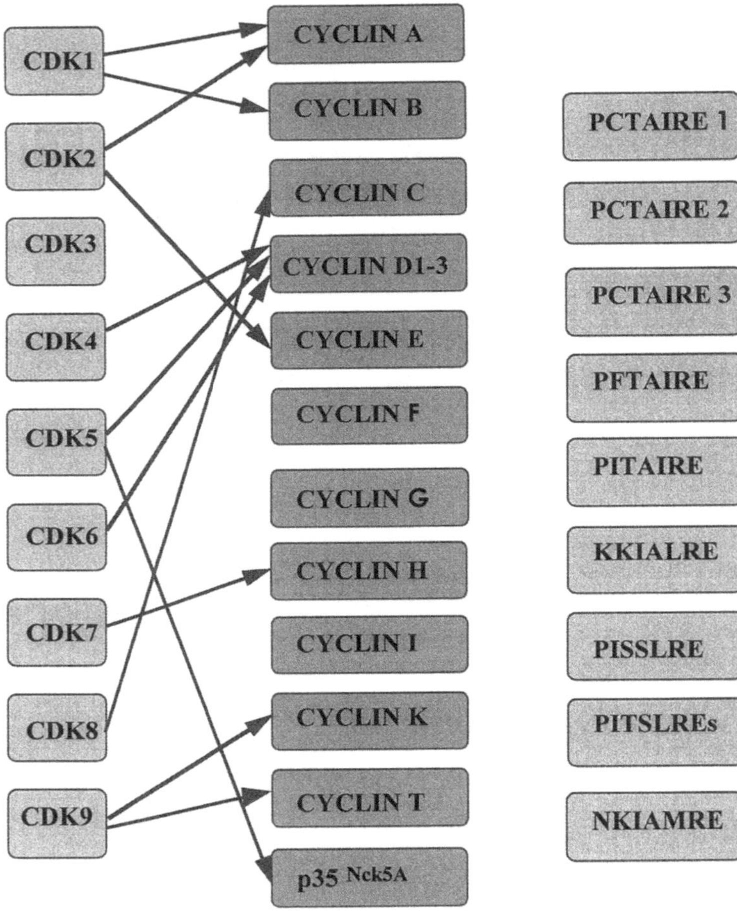

Fig. 1. CDKs, cyclins and CDK-related kinases. Arrows link cyclins to CDKs as found in vivo

the first member of a large family of kinase regulators (Fig. 1). The availability of the complete genomic sequence of *Caenorhabditis elegans* provides the first complete vision of the extent of the CDK family in an eucaryote (see overview in [7]): among the 19,099 predicted genes, 2.6% encode protein kinases. Among these 493 kinases, there are 14 CDKs and the genome encodes 34 cyclins.

4.1.1 CDKs and Related Kinases: Structure

CDKs are typical Ser/Thr kinases (about 300 amino acids, molecular weight: 33–40 kDa) which display the eleven subdomains shared by all protein kinases (see the protein kinase resource site: http://www.sdsc.edu/kinases). Nine CDKs and eleven cyclins have been identified in man: the known CDK/cyclin complexes are presented in Fig. 1. The CDKs which associate with cyclin F, G and I have not been identified yet. In addition, there are several "CDK-related kinases" with no identified cyclin partner (Fig. 1). These are easily recognised by their sequence homology to *bona fide* CDKs and by the presence of a variation of the conserved "PSTAIRE" motif, located in the cyclin-binding domain (sub-domain III) [8]. Until their associated cyclin is discovered (if any is associated), these "CDK-related kinases" are named following the sequence of their PSTAIRE motif: PCTAIRE 1–3, PFTAIRE, PITAIRE, KKIALRE, PISSLRE, NKIAMRE and the PITSLRE. To be fully active, CDK/cyclin complexes have to be phosphorylated on the residue corresponding to CDK2 Thr160, located on the T-loop of the kinase. This phosphorylation is carried out by CDK7/cyclin H in association with a third protein, MAT1. The CDK subunit must also be dephosphorylated on Thr14 and Tyr15, two residues located at the border of the ATP-binding pocket.

The structure of CDK2 consists of an amino-terminal lobe (residues 1–85) rich in β-sheets and a larger, mostly α-helical, carboxy-terminal lobe. The ATP binding site is located in a deep cleft between the two lobes which contains the catalytic residues conserved among eucaryotic protein kinases. Crystallographic studies, performed using monomeric CDK2, and CDK2/cyclin A, have shown the important influence cyclin binding has on CDK2 structure. Cyclin A binding induces important changes in the CDK2 structure, forcing the kinase into an active conformation. First, the T-loop, which blocks the substrate access in monomeric CDK2, is found outside the catalytic cleft exposed to the solvent after cyclin A binds. This conformational change of the T-loop also allows the activating phosphorylation on Thr160. The second conformational change induced by cyclin binding is found within the ATP-binding site where a reorientation of the amino acids side chains induces the alignment of ATP's triphosphate necessary for phosphate transfer. The strong sequence homology between the catalytic domains of different CDKs suggests that their tridimensional structures will be similar.

4.1.2 CDKs and Related Kinases: Functions

4.1.2.1 CDKs and Cell Cycle Control

The adult human being is constituted of 10^{13} cells which are all derived from the initial fertilised egg. Every second our body undergoes 20 million cell divisions to compensate for continuous cell loss and death. Progression through the G1, S, G2 and M phases of the cell division cycle is directly controlled by the transient activation of various CDKs. In early to mid G1, extracellular signals modulate the activation of a first set of CDKs, CDK4 and CDK6 associated with D-type cyclins. CDK4/cyclin D1 and CDK6/cyclin D3 phosphorylate the retinoblastoma protein pRb and other members of the pRb family. Phosphorylation inactivates pRb, resulting in the release of the E2F and DP1 transcription factors which, in turn, control the expression of genes whose products are required for the G1/S transition and S phase progression, such as CDK2, cyclin E and cyclin A. The CDK2/cyclin E complex which is responsible for the G1/S transition also causes further phosphorylation of pRb allowing the release of an increased amount of transcription factors. During S phase, CDK2/cyclin A phosphorylates different substrates allowing DNA replication and the inactivation of the G1 transcription factors. Around the time of the S/G2 transition, CDK1 associates with cyclin A. Slightly later, CDK1/cyclin B appears and triggers the G2/M transition by phosphorylating a large set of substrates such as the nuclear lamins. Phosphorylation of APC, the "Anaphase Promoting Complex", by CDK1/cyclin B is required for cyclin B proteolysis, transition to anaphase and completion of mitosis. These successive waves of CDK/cyclin assemblies and activations are tightly regulated by post-translational modifications and intracellular translocations. They are co-ordinated and dependent on the completion of previous steps, through so-called "checkpoint" controls (review in ref. [1–4]).

4.1.2.2 CDKs and Transcription

Beside their roles in controlling the cell cycle, some CDKs directly influence transcription. The CDK7/cyclin H/MAT1 complex is a component of the TFIIH complex, a basal transcription factor. TFIIH kinase activity is responsible for phosphorylation of the C-terminal domain of the large subunit of RNA polymerase II (CTD RNA pol II), required for the elongation process.

CDK8 associates with cyclin C and has been found in a multiprotein complex with RNA polymerase II. Like CDK7/cyclin H, CDK8/cyclin C phosphorylates CTD RNA pol II, but on different sites, suggesting a distinct mechanism of transcriptional regulation.

CDK9/cyclin T is a component of the positive transcription elongation factor P-TEFb. It is responsible for the Tat-associated kinase activity involved in the HIV-1 Tat transactivation. It also displays CTD RNA pol II kinase activity.

4.1.2.3 CDKs and Neural and Muscular Functions

CDK5 has been purified from bovine brain where it associates with cytoskeletal proteins, such as the tau protein and the neurofilaments NF-H and NF-M. CDK5 activity is important for outgrowth of neurites and neuronal development. CDK5 also plays a crucial role in myogenesis and somites organisation in *Xenopus* embryos and in remodelling tissues. There is a clear involvement of CDK5 in the apoptotic process, as illustrated by a positive correlation between the activity of CDK5 and the number of cells undergoing apoptosis, in both developmental and remodelling tissues.

Another interesting aspect of CDK5 is the nature of its associated regulatory subunits, p35 or p25, a 25 kDa protein derived by proteolytic clivage from the 35 kDa precursor. Despite their evolutionary distance from cyclins, these proteins function as CDK5 activators in place of the classical cyclins. Nevertheless the predicted structure of p35 shows a similar fold to that of cyclins, which explains the efficient activation of CDK5 and also extends the list of potential activators for CDK-related proteins. It was recently shown that conversion of p35 to p25 leads to constitutive activation of CDK5, and alteration of its cellular localisation and substrate specificity [9]. CDK5/p25 expression in cultured primary neurons triggers apoptosis [9]. These findings, as well as the accumulation of p25 [9] and increased CDK5 activity in Alzheimer's disease patients' brains, indicate that CDK5 activation may be involved in the cytoskeletal abnormalities and neuronal death observed in this neurodegenerative disorder.

Finally CDK5 was recently demonstrated as a downstream element of dopamine signaling [10]. When phosphorylated on Thr34 by PKA, the striatum-specific DARPP-32 protein is an inhibitor of phosphatase 1; when phosphorylated on Thr75 by CDK5/p25, DARPP-32 becomes an inhibitor of PKA. In vivo phosphorylation on this site does not occur in $p35^{-/-}$ tissue [10].

4.1.2.4 CDKs and Apoptosis

In addition to a possible role of CDK5 in neuronal cell death, other enzymes of this family may be involved in apoptosis. The PITSLRE family of CDK-related kinases contains more than 20 isoforms derived from three different genes and alternative splicing. A caspase-dependent proteolytic cleavage in the N-terminal region of some of these isoforms leads to a 50 kDa active kinase involved in apoptosis. It has been recently demonstrated that CDK2, in association with an unidentified protein different from cyclin A or E, is upregulated in thymocytes undergoing apoptosis. This CDK activity is required for induction of apoptosis, providing a very interesting link between cell division and cell death (see also review in [11]).

4.2 ATP Site-Directed CDK Inhibitors

The observation that CDK activity becomes abnormally regulated in human cancers and the direct involvement of CDK5/p25 in Alzheimer's disease have stimulated the search for chemical inhibitors of these kinases (see reviews in [12–16]; see also KID, the Kinase Inhibitor Database: http://www.sb-roscoff.fr/KID). The first identified molecules display anti-proliferative and apoptosis-inducing properties and are being evaluated as potential anti-tumour agents. They also inhibit in vivo the hyperphosphorylation of the tau protein, a diagnostic feature of Alzheimer's disease. Other therapeutic applications are also under study. All inhibitors identified so far act by competitive inhibition of ATP binding.

4.2.1 CDK Inhibitors: Chemical Variety, Efficiency, Selectivity

The search for CDK inhibitors has mostly been based on the use of CDK1/ cyclin B as a molecular target. Starfish oocytes have become a widely used source of purified enzyme [17]. Alternatively, recombinant CDKs have been expressed in insect cells and used to screen for inhibitors. The purified CDKs are assayed with ^{32}P-γ-ATP and an appropriate protein substrate such as histone H1 or the retinoblastoma protein in the presence of an increasing concentration of potential inhibitors. The dose-response curves provide IC_{50} values which are currently used to compare the efficiency of compounds to one another. Using these methods, eleven specific inhibitors have been identified (Table 1): olomoucine [18], roscovitine [19, 20], purvalanol [21, 22], CVT-313 [23], toyocamycin [24], flavopiridol [25], CGP60474 [26], indirubin-3'-monoxime [27], the paullones [28, 29], γ-butyrolactone [30] and hymenialdisine [31]. They all derive from structure/activity studies and from molecular modelling based on the crystal structure of the inhibitor in complex with CDK2. Despite their chemical variety, these inhibitors all act by competing with ATP at the ATP-binding site of the catalytic subunit of the kinase.

The selectivity of some of these inhibitors is usually quite remarkable [11–16]. Some inhibit CDK1, CDK2 and CDK5 but have no effect on CDK4 and CDK6. The structural reasons for such selectivity are unknown. No CDK4/CDK6 selective inhibitor has been reported yet. A few less selective inhibitors have been described ([12]): 6-dimethylaminopurine, isopentenyladenine, suramin, staurosporine, UCN-01, 9-hydroxyellipticine. Although their use as tools in cell biology is limited, this is not necessarily the case in therapy. Furthermore they may constitute the basis for identification of more selective inhibitors as illustrated by olomoucine, roscovitine, purvalanol, which are all derived from the non-selective kinase inhibitors 6-dimethylaminopurine and isopentenyladenine.

Table 1. Chemical inhibitors of CDKs: IC_{50} towards CDK1, availability of a CDK2 co-crystal structure and selectivity

Inhibitor	IC_{50} (μM) on CDK1/cyclin B	Crystal Structure with CDK2	Selectivity
6-Dimethylaminopurine	120.000	no	poor
Isopentenyladenine	55.000	[66]	poor
Olomoucine [18]	7.000	[66]	++++
Roscovitine [19, 20]	0.450	[20]	++++
CVT-313 [23]	4.200	no	unknown
Purvalanol A&B [21, 22]	0.004	[21]	++++
Flavopiridol [25]	0.400	[67] des-chloro-flavopiridol	++
Suramin	4.000	no	poor
9-hydroxyellipticine	1.000	no	poor
Toyocamycin [24]	0.880	no	unknown
Staurosporine	0.004	[68]	poor
γ-Butyrolactone [30]	0.600	no	+++
CGP60474 [26]	0.020	no	unknown
Kenpaullone [29]	0.400	no	++++
Alsterpaullone [28]	0.035	no	++++
Indirubin-3'-monoxime [27]	0.180	[27]	+++
Hymenialdisine [31]	0.022	[31]	++

4.2.2 CDK2/Inhibitor Crystal Structures

Co-crystallisation with CDK2 has allowed spectacular progress in our comprehension of the mechanism of action of the inhibitors (review in [15]). Besides ATP, eight compounds have been crystallised with CDK2 (Table 1). The high resolution crystal structures revealed that all inhibitors bind in the deep groove located between the N and C-terminal domains of the kinase and normally occupied by the adenine ring of ATP. The planar heterocyclic ring system, a structural feature common to most CDK inhibitors, in each case occupies a hydrophobic cleft surrounded by Ile 10, Val 18, Phe 82, and Leu 134. The Ile10, Leu 83 and Leu134 residues are responsible for 40% of all contacts between CDK2 and the inhibitors. Two or three hydrogen bonds are constantly found between the inhibitors and the Glu81 and Leu83 residues of

CDK2. The selectivity probably comes from interactions with residues, conserved in CDKs but not in other kinases, which do not interact with ATP. The ATP-binding site of CDK2 has thus an impressive capacity to accommodate a variety of inhibitors containing flat heterocyclic structures. Although all the inhibitors overlap significantly within the ATP site each structure provides information that can potentially be used to improve the affinity or alter the specificity of the other inhibitors. Continued screening of compound collections, both synthetic and natural, will undoubtedly yield compounds that bind even more efficiently to the ATP binding site.

4.2.3 CDK Inhibitors: Cellular Effects

Despite great *in vitro* selectivity, the *in vivo* selectivity of CDK inhibitors remains an open question. Numerous factors affect the cellular effects of an inhibitor: membrane permeability, accumulation in various cell compartments, metabolism into inactive compounds, competition with high intracellular ATP concentrations, binding to other non identified targets, ...etc., probably a combination of all these factors. For these reasons, in vivo results obtained with kinase inhibitors should be interpreted with caution. It is important to show that CDK substrates are not phosphorylated anymore following cell treatment with the inhibitory compound. Inhibitors belonging to different chemical classes should have similar effects. The *in vivo* efficiency of a series of chemically related inhibitors should parallel the *in vitro* efficiency. Under these conditions, the use of chemical inhibitors of CDKs will nicely complement, in fundamental cell biology, the use of dominant negatives, the overexpression of natural CDK inhibitors (CKI), the micro-injection of inactivating antibodies or antisense RNAs.

A large number of experiments have been performed on a wide variety of cellular models confirming that CDK inhibitors have anti-proliferative properties (review in [11]). In general they inhibit cell proliferation both in G1 and at the end of G2/prophase. These results support the idea that they act *in vivo* on cell proliferation through inhibition of both CDK2 and CDK1.

Induction of apoptosis is often associated with cell cycle arrest. There are three types of effects of CDK inhibitors on apoptosis:

1) lack of effect: Myc-induced apoptosis is insensitive to CDK inhibitors.
2) induction of apoptosis: CDK inhibitors either trigger cell death or facilitate apoptosis triggered by other agents. This is the case for most proliferating cells.
3) inhibition of apoptosis: such is the case for differentiated neuronal cells and thymocytes. CDK inhibitors protect these cells from apoptosis triggered by various factors.

4.2.4 CDK Inhibitors: Future Prospects

4.2.4.1 The Real Targets of CDK Inhibitors

We anticipate that more active inhibitors will soon come out of the large number of screening programmes developed throughout the world. Although some families of compounds have provided very selective leads, our ability to assess selectivity is limited by the number of enzymes we can test.

We anticipate the development of two general methods for the identification of the cellular targets of active compounds: gene expression arrays and affinity chromatography.

In the first approach, RNA is extracted from control and CDK inhibitor-treated cells, labelled and then hybridized to a membrane where cDNAs fragments corresponding to genes involved in various cellular events (apoptosis, cell cycle,...etc.) are arrayed. This method provides a wide profile of gene expression following treatment with a drug. We believe that this method will soon be used in understanding the mechanisms of action of CDK inhibitors. A similar method, involving hybridization to high density oligonucleotide arrays (Affymetrix technology), was used recently to follow the expression pattern of the whole *S. cerevisiae* genome following treatment with flavopiridol and a purine inhibitor [21].

The second approach is initially based on the crystal structure of CDK/inhibitor complexes. This data allows visualisation of the inhibitor orientation within the ATP-binding pocket. It tells us which part of the inhibitor is suitable for addition of a linker which can be used to tether the inhibitor to a solid matrix. We have recently developed this approach with purine derivatives [32, 33]. The immobilised purine is used as a chromatography media to purify interacting proteins from cell extracts. As an example we have used starfish prophase and metaphase oocytes to investigate the nature of the targets of purvalanol. The immobilised purvalanol affinity reagent allows the recovery of CDK1/cyclin B. In addition to CDK1 and cyclin B, a very small number of proteins (as detected by silver staining SDS-polyacrylamide gels) interact with the resin. Following protein microsequencing, the two major ones have been identified as calmodulin-dependent kinase 2 and ribosomal S6 kinase [33]. This chromatography approach can be applied with any new inhibitor to identify its potential cellular targets. We anticipate that this method will develop considerably and will complement the gene expression studies.

4.2.4.2 The Potential Applications of CDK Inhibitors

Only a few years have elapsed from the initial discovery of cdc2 in *Schizosaccharomyces pombe* by Nurse and Bissett [5] to the use of CDK1/cyclin B as an enzymatic screening target [17], the identification of several potent inhibitors (Table 1) and the first clinical trials [34].

Although the search for CDK inhibitors was initially started because of their anti-tumour potential, many new applications are now being investigated (reviews in [15, 16]). Indeed several diseases derive from the unbalanced but non-tumoral proliferation of certain cells, such as atherosclerosis, post-angioplastic restenosis, tumour angiogenesis, glomerulonephritis (kidney glomerula cell proliferation), psoriasis (skin fibroblast proliferation). Numerous viruses require active CDKs for their replication, some actually encoding modified cyclins. There are several examples where CDK inhibitors are found to efficiently restrain viral replication: human cytomegalovirus, herpes virus, varicella-zoster virus. The HIV Tat protein activates a kinase recently identified as CDK9/cyclin T. CDK inhibitors may also have major applications in the control of neurodegenerative disorders. Indeed CDK5/p35 appears to be one of the kinases which hyperphosphorylates the microtubule-associated protein tau in Alzheimer's disease (review in ref. [35]). This hyperphosphorylation is responsible for the appearance of the paired helical filaments (PHF) typically associated with Alzheimer's disease. CDK inhibitors appear to protect neuronal cells from apoptosis, although the underlying mechanisms are unclear. Altogether this data strongly encourage the study of CDK inhibitors as potential pharmacological agents to control neurodegeneration. Another potential application currently being investigated is the use of CDK inhibitors and their derivatives against eukaryotic parasitic organisms *Plasmodium*, *Trypanosoma*, *Leishmania*, *Toxoplasma*, microsporidia,...etc.). The multiplication and development of such parasites are likely to depend on the activity of CDKs or CDK-related kinases (CRKs) (review in [36] for *Plasmodium* kinases). Homology of these proteins with CDKs found in vertebrates (usually 40–60% identity) is sufficient to allow their classification as CDKs or CRKs, but leaves room for significant divergences. This opens a potentially promising therapeutic window, as there is a possibility that structural differences between parasite and host CDKs may result in differential affinities for inhibitory molecules. Targeting the ATP-binding pocket of CDKs presents the advantage that this part of the protein is unlikely to allow many variations, making resistance less likely to emerge than in the case of more variable targets.

4.3 CDKs and Cyclins in the Testis

A few studies have focused on CDKs and cyclins in the testis. They are briefly reviewed below.

4.3.1 CDKs and Related Kinases

The expression of CDK1, CDK2, CDK4, CDK5, PCTAIRE-1 and PCTAIRE-3 has been followed during murine germ cell development [37–39]. CDKs are expressed in meiotic spermatocytes, but also in non-proliferating cells such as Sertoli cells [37]. CDK5 is expressed in Leydig and Sertoli cells [40]. GAK, an association partner of cyclin G and CDK5 is expressed ubiquitously, with the highest level of expression in the testis [41]. The activity of CDK1 in the testis requires the presence of the HSP70–2 molecular chaperone [42]. PCTAIRE-1 and –3 is mostly expressed in postmeiotic spermatids [37]. PFTAIRE-1 [38] is highly expressed in brain, late pachytene spermatocytes and embryo. These results suggest that these kinases may be involved in gametogenesis but also in other regulatory processes, such as differentiation.

4.3.2 Cyclins

The expression and functions of cyclins has been extensively investigated in the testis.

Cyclin A1 is a recently cloned cyclin [43, 44] with exclusive expression in meiotic cells in the testis (and in acute myeloid leukemia). Cyclin A2 is expressed in somatic tissues [43] but also in preleptotene spermatocytes [45, 46]. The levels of cyclin A1 mRNA rise in late spermatocytes and become undetectable soon after the completion of meiotic divisions [43]. Its promoter region was recently cloned [47]. Cyclin A2 expression decreases, while cyclin G1 increases as Leydig cells withdraw from the cell cycle during pubertal development [48]. Cyclin A1 associates with CDK1 and CDK2 while cyclin A2 only associates with CDK2 [46]. Cyclin A1(-/-) male mice are sterile due to a block of spermatogenesis before the first meiotic division, whereas females are normal [49]. Meiosis arrest is associated with an increase in germ cell apoptosis [49].

Cyclin B1 (along with CDK1) is expressed in male germ cells [50, 51], with highest levels in pachytene spermatocytes and round spermatids, and lowest levels in postmeiotic spermatids [52, 53]. Cyclin B1(-/-) mice die in utero, while Cyclin B2(-/-) mice develop normally and are fertile [54], despite the high levels and distinct expression patterns of cyclin B2 during spermatogenesis [50]. Cyclin B1 may thus compensate for the absence of cyclin B2, but the reverse is not true.

The expression patterns of cyclins D1, D2 and D3 during testicular development in the mouse are quite distinct [55]: cyclin D1 is expressed in the non-dividing Sertoli cells, cyclin D2 level is slightly enriched in germ cell-deficient testes, cyclin D3 is highest in non-dividing, haploid, round spermatids and in the cytoplasm of spermatocytes [56]. Cyclin D2(-/-) male mice display hypoplastic testes [57]. Some human testicular tumours express high levels of cyclin D2 [57, 58]. The expression of D-type cyclins in adult human testis and testicular cancer has been described by Bartkova et al. [59]. The pattern is dominated by high levels of cyclin D3 in quiescent Leydig cells and the lack of any D-type cyclins in the germ cells. Cyclin D3 is also developmentally regulated, its associated kinase activity is only detectable in immature testes [60].

An isoform of cyclin I is mostly expressed in human testis [61].

4.3.3 CDK Inhibitors

CDKs are regulated by several families of inhibitory proteins (review in [3]). The expression of some of these proteins has been investigated in mammalian testis: $p21^{cip1/WAF1}$ is expressed in pachytene spermatocytes up to spermatids [62, 63]. X-irradiation increases the level of p21. $p27^{Kip1}$ is expressed mostly in the quiescent cells of the testis (terminally differentiated Sertoli cells, 30% of the Leydig cells mice) [64]. Aberrations of the spermatogenic process are observed in p27(-/-) mice [64]. The CDK4/CDK6 inhibitor $p19^{INK4d}$ has been recently knocked out in mice [65]: males exhibit a marked testicular atrophy associated with an increased apoptosis of germ cells, although they remain fertile [65].

4.4 Conclusions

Since their discovery the functions of CDKs in cell cycle regulation have been extensively investigated in a large variety of models with diverse techniques. In this context the regulation of meiotic divisions in male germ cells clearly deserves more studies. The expression of CDKs and related kinases as well as cyclins in non-proliferative cells in the testis clearly suggests an involvement of these regulators in other functions such as apoptosis and differentiation. These aspects also deserve more investigation. The recently developed molecular biology and pharmacological tools should provide an appreciable help in the study and understanding of these functions.

Acknowledgements. We are grateful to Armelle Jezequel for secretarial help. This work was supported by grants from the "Association pour la Recherche sur le Cancer" (ARC 9314) and the "Conseil Régional de Bretagne".

References

1. Morgan D (1997) Cyclin-dependent kinases: engines, clocks, and microprocessors. Annu Rev Cell Dev Biol 13: 261–291
2. Meijer L, Guidet S, Philippe M (editors) (1997) Progress in Cell Cycle Research, vol. 3, Plenum Press, New York, 321 pp (24 chapters)
3. Vogt PK, Reed SI (1998) Cyclin dependent kinase (CDK) inhibitors. Current Topics in Microbiology and Immunology, Springer Verlag, 169 pp
4. Meijer L, Jezequel A Ducommun B (editors) (2000). Progress in Cell Cycle Research, vol. 4, Plenum Press, New York, 248 pp (21 chapters)
5. Nurse P, Bissett Y (1981) Gene required in G1 for commitment to cell cycle and in G2 for control of mitosis in fission yeast. Nature 292: 558–60
6. Lee MG, Nurse P (1987) Complementation used to clone a human homologue of the fission yeast cell cycle control gene cdc2. Nature 237: 31–35
7. Plowman GD, Sudarsanam S, Bingham J, Whyte D and Hunter T (1999) The protein kinases of Caenorhabditis elegans: a model for signal transduction in multicellular organisms. Proc Natl Acad Sci USA 96: 13603–13610
8. Meyerson M, Enders GH, Wu CL, Su LK, Gorka C, Nelson C, Harlow E, Tsai LH (1992) A family of human cdc2-related protein kinases. EMBO J 11: 2909–2917
9. Patrick GN, Zukerberg L, Nikolic M, De la Monte S, Dikkes P, Tsai LH (1999) Conversion of p35 to p25 deregulates Cdk5 activity and promotes neurodegeneration. Nature 402: 615–622
10. Bibb JA, Snyder GL, Nishi A, Yan Z, Meijer L, Fienberg AA, Tsai LH, Kwon YT, Girault J-A, Czernik AJ, Huganir RL, Hemmings HC, Nairn AC, Greengard P (1999) Phosphorylation of DARPP-32 by Cdk5 modulates dopamine signalling in neurons. Nature 402: 669–671
11. Guo M, Hay BA (1999) Cell proliferation and apoptosis. Curr Opin Cell Biol 11: 745–752
12. Meijer L (1996) Chemical inhibitors of cyclin-dependent kinases. Trends in Cell Biol 6: 393–397
13. Meijer L, Kim SH (1997) Chemical inhibitors of cyclin-dependent kinases. Methods in Enzymology, "Cell Cycle Control", vol. 283: 113–128
14. Garrett MD, Fattaey A (1999) Cyclin-dependent kinase inhibition and cancer therapy. Curr Opin Genet & Dev 9: 104–111
15. Gray N, Détivaud L, Doerig C, Meijer L (1999) ATP-site directed inhibitors of cyclin-dependent kinases. Curr Medicin Chem 6, 859–876
16. Meijer L, Leclerc S, Leost M (1999) Properties and potential applications of chemical inhibitors of cyclin-dependent kinases. Pharmacol & Ther 82: 279–284

17. Rialet V, Meijer L (1991) A new screening test for antimitotic compounds using the universal M phase-specific protein kinase, p34^{cdc2}/cyclin B^{cdc13}, affinity-immobilized on p13^{suc1}-coated microtitration plates. Anticancer Res 11: 1581–1590
18. Vesely J, Havlicek L, Strnad M, Blow JJ, Donella-Deana A, Pinna L, Letham D S, Kato JY, Détivaud L, Leclerc S, Meijer L (1994) Inhibition of cyclin-dependent kinases by purine derivatives. Eur J Biochem 224: 771–86
19. Meijer L, Borgne A, Mulner O, Chong JPJ, Blow JJ, Inagaki N, Inagaki M, Delcros JG, Moulinoux JP (1997) Biochemical and cellular effects of roscovitine, a potent and selective inhibitor of the cyclin-dependent kinases cdc2, cdk2 and cdk5. Eur J Biochem 243: 527–36
20. de Azevedo WF, Leclerc S, Meijer L, Havlicek L, Strnad M, Kim S-H (1997) Inhibition of cyclin-dependent kinases by purine analogues: crystal structure of human cdk2 complexed with roscovitine. Eur J Biochem 243: 518–26
21. Gray N, Wodicka L, Thunnissen AM, Norman T, Kwon S, Espinoza FH, Morgan DO, Barnes G, Leclerc S, Meijer L, Kim SH, Lockhart DJ, Schultze P (1998) Exploiting chemical libraries, structure, and genomics in the search for new kinase inhibitors. Science 281: 533–538
22. Chang YT, Gray NG, Rosania GR, Sutherlin DP, Kwon S, Norman TC, Sarohia R, Leost M, Meijer L, Schultz PG (1999) Synthesis and application of functionally diverse 2,6,9-trisubstituted purine libraries as CDK inhibitors. Chem & Biol 6: 361–375
23. Brooks EE, Gray NS, Joly A, Kerwar SS, Lum R, Mackman RL, Norman TC, Rosete J, Rowe M, Schow SR, Schultz PG, Wang X, Wick MM, Shiffman D (1997) CVT-313, a specific and potent inhibitor of CDK2 that prevents neointimal proliferation. J Biol Chem 272: 29207–29211
24. Park SG, Cheon JY, Lee YH, Park J–S, Lee KY, Lee CH, Lee SK (1996) A specific inhibitor of cyclin-dependent protein kinases, CDC2 and CDK2. Mol Cells 6: 679–83
25. Sedlacek HH, Czech J, Naik R, Kaur G, Worland P, Losiewicz M, Parker B, Carlson B, Smith A, Senderowicz A, Sausville E (1996) Flavopiridol (L86 8275; NSC 649890), a new kinase inhibitor for tumor therapy. Internat J Oncol 9: 1143–1168
26. Zimmermann J (1995) Pharmacologically active pyrimidine derivatives and processes for the preparation thereof. PCT Ciba-Geigy, WO 95/09853
27. Hoessel R, Leclerc S, Endicott J, Noble M, Lawrie A, Tunnah P, Leost M, Damiens E, Marie D, Marko D, Niederberger E, Tang W, Eisenbrand G, Meijer L (1999) Indirubin, the active constituent of a Chinese antileukaemia medicine, inhibits cyclin-dependent kinases. Nature Cell Biol 1, 60–67

28. Schultz C, Link A, Leost M, Zaharevitz DW, Gussio R, Sausville EA, Meijer L, Kunick C (1999) The paullones, a series of cyclin-dependent kinase inhibitors: synthesis, evaluation of CDK1/cyclin B inhibition, and *in vitro* antitumor activity. J Med Chem 42: 2909–2919
29. Zaharevitz D, Gussio R, Leost M, Senderowicz AM, Lahusen T, Kunick C, Meijer, L, Sausville EA (1999) Discovery and initial characterization of the paullones, a novel class of small-molecule inhibitors of cyclin-dependent kinases. Cancer Res 59: 2566–2569
30. Kitagawa M, Okabe T, Ogino H, Matsumoto H, Suzuki-Takahashi I, Kokubo T, Higashi H, Saitoh S, Taya Y, Yasuda H, Ohba Y, Nishimura S, Tanaka N, Okuyama A (1993) Butyrolactone I, a selective inhibitor of cdk2 and cdc2 kinase. Oncogene 8: 2425–2432
31. Meijer L, Thunissen amwh, White A, Garnier M, Nikolic M, Tsai LH, Walter J, Cleverley KE, Salinas PC, Wu YZ, Biernat J, Mandelkow EM, Kim S-H, Pettit GR (2000) Inhibition of cyclin-dependent kinases, GSK-3β and casein kinase 1 by hymenialdisine, a marine sponge constituent. Chem & Biol 7: 51–63
32. Rosania GR, Merlie J, Gray NS, Chang YT, Schultz PG, Heald R (1999) A cyclin-dependent kinase inhibitor inducing cancer cell differentiation: biochemical identification using Xenopus egg extracts. Proc Natl. Acad Sci USA 96: 4797–4802
33. Knockaert M, Gray NS, Damiens E, Chang YT, Grellier P, Grant K, Fergusson D, Mottram J, Soete M, Dubremetz JF, LeRoch K, Doerig C, Schultz PG, Meijer L (2000) Intracellular targets of cyclin-dependent kinase inhibitors: identification by affinity chromatography using immobilised ligants. Chem & Biol, in press.
34. Senderowicz A, Headlee D, Stinson SF, Lush RM, Kalil N, Villalba L, Hill K, Steinberg SM, Figg WD, Tompkins A, Arbuck SG, Sausville EA (1998) Phase I trial of continuous infusion flavopiridol, a novel cylin-dependent kinase inhibitor in patients with refractory neoplasms. Clin Oncol 16: 2986–2999
35. Mandelkow E (1999) The tangled tale of tau. Nature 402, 588–589
36. Kappes B, Doerig CD, Graeser R (1999) An overview of Plasmodium protein kinases. Parasitology Today 15: 449–454
37. Rhee K, Wolgemuth DJ (1995) Cdk family genes are expressed not only in dividing but also in terminally differentiated mouse germ cells, suggesting their possible function during both cell division and differentiation. Dev Dyn 204: 406–420
38. Besset V, Rhee K, Wolgemuth DJ (1998) The identification and characterization of expression of Pftaire-1, a novel Cdk family member, suggest its function in the mouse testis and nervous system. Mol Reprod Dev 50: 18–29

39. Besset V, Rhee K, Wolgemuth DJ (1999) The cellular distribution and kinase activity of the Cdk family member Pctaire1 in the adult mouse brain and testis suggest functions in differentiation. Cell Growth Differ 10: 173–181
40. Musa FR, Tokuda M, Kuwata Y, Ogawa K, Tomizawa k, Konishi R, Takenaka I, Hatase O, (1998) Expression of cyclin-dependent kinase 5 and associated cyclins in Leydig and Sertoli cells of the testis. J Androl 19: 657–666
41. Kimura SH, Tsuruga H, Yabuta N, Endo Y, Nojima H (1997) Structure, expression, and chromosomal localization of human GAK. Genomics 44: 179–187
42. Zhu D, Dix DJ, Eddy EM (1997) HSP70–2 is required for CDC2 kinase activity in meiosis I of mouse spermatocytes. Development 124(15): 3007–3014
43. Sweeney C, Murphy M, Kubelka M, Ravnik SE, Hawkins CF, Wolgemuth DJ, Carrington M (1996) A distinct cyclin A is expressed in germ cells in the mouse. Development 122: 53–64
44. Yang R, Morosetti R, Koeffler HP (1997) Characterization of a second human cyclin A that is highly expressed in testis and in several leukemic cell lines. Cancer Res 57: 913–920
45. Ravnik SE, Wolgemuth DJ (1996) The developmentally restricted pattern of expression in the male germ line of a murine cyclin A, cyclin A2, suggests roles in both mitotic and meiotic cell cycles. Dev Biol 173: 69–78
46. Ravnik SE, Wolgemuth DJ (1999) Regulation of meiosis during mammalian spermatogenesis: the A-type cyclins and their associated cyclin-dependent kinases are differentially expressed in the germ-cell lineage. Dev Biol 207: 408–418
47. Muller C, Yang R, Beck-von-Peccoz L, Idos G, Verbeek W, Koeffler HP (1999) Cloning of the cyclin A1 genomic structure and characterization of the promoter region. GC boxes are essential for cell cycle-regulated transcription of the cyclin A1 gene. J Biol Chem, 274:11220–11228
48. Ge RS, Hardy MP (1997) Decreased cyclin A2 and increased cyclin G1 levels coincide with loss of proliferative capacity in rat Leydig cells during pubertal development. Endocrinology 138: 3719–3726
49. Liu D, Matzuk MM, Sung WK, Guo Q, Wang P, Wolgemuth DJ (1998) Cyclin A1 is required for meiosis in the male mouse. Nat Genet 20: 377–380
50. Chapman DL, Wolgemuth DJ (1992) Identification of a mouse B-type cyclin which exhibits developmentally regulated expression in the germ line. Mol Reprod Dev 33: 259–269
51. Chapman DL, Wolgemuth DJ (1993) Isolation of murine cyclin B2 cDNA and characterization of the lineage and temporal specificity of expression

of the B1 and B2 cyclins during oogenesis, spermatogenesis and early embryogenesis. Development 118: 229–240
52. Chapman DL, Wolgemuth DJ (1994) Regulation of M-phase promoting factor activity during development of mouse male germ cells. Dev Biol 165: 500–506
53. Gromoll J, Wessels J, Rosiepen G, Brinkworth MH, Weinbauer GF (1997) Expression of mitotic cyclin B1 is not confined to proliferating cells in the rat testis. Biol Reprod 57: 1312–1319
54. Brandeis M, Rosewell I, Carrington M, Crompton T, Jacobs Ma, Kirk J, Gannon J, Hunt T (1998) Cyclin B2-null mice develop normally and are fertile whereas cyclin B1-null mice die in utero. Proc Natl Acad Sci USA 95: 4344–4349
55. Ravnik SE, Rhee K, Wolgemuth DJ (1995) Distinct patterns of expression of the D-type cyclins during testicular development in the mouse. Dev Genet 16: 171–178
56. Kang MJ, Kim MK, Terhune A, Park JK, Kim YH, Koh GY (1997) Cytoplasmic localization of cyclin D3 in seminiferous tubules during testicular development. Exp Cell Res 234: 27–36
57. Sicinski P, Donaher JL, Geng Y, Parker SB, Gardner H, Park MY, Robker RL, Richards JS, McGinnis LK, Biggers JD, Eppig JJ, Bronson RT, Elledge SJ, Weinberg RA (1996) Cyclin D2 is an FSH-responsive gene involved in gonadal cell proliferation and oncogenesis. Nature 384: 470–474
58. Houldsworth J, Reuter V, Bosl GJ, Chagantis RS (1997) Aberrant expression of cyclin D2 is an early event human male germ cell tumorigenesis. Cell Growth Differ 8: 293–299
59. Bartkova J, Rajpert-de Meyts E, Skakkebak NE, Bartek J (1999) D-type cyclins in adult human testis and testicular cancer: relation to cell type, proliferation, differentiation, and malignancy. J Pathol 187: 573–581
60. Zhang Q, Wang X, Wolgemuth DJ (1999) Developmentally regulated expression of cyclin D3 and its potential in vivo interacting proteins during murine gametogenesis. Endocrinology 140: 2790–2800
61. Zhu X, Naz RK (1998) Expression of a novel isoform of cyclin I in human testis. Biochem Biophys Res Commun 249: 56–60
62. Beumer TL, Roepers-Gajadien HL, Gademan IS, Rutgers DH, de Rooij DG (1997) P21(Cip/WAF1) expression in the mouse testis before and after irradiation. Mol Reprod Dev 47: 240–247
63. West A, Lahdetie J (1997) p21WAF1 expression during spermatogenesis of the normal and X-irradiated rat. Int J Radiat Biol 71: 283–291
64. Beumer TL, Kiyokawa H, Roepers-Gajadien HL, van den Bos LA, Lock TM, Gademan IS, Rutgers DH, Koff A, de Rooij DG (1999) Regulatory role of p27kip1 in the mouse and human testis. Endocrinology 140: 1834–1840

65. Zindy F, van Deursen J, Grosveld G, Sherr CJ, Roussel MF (2000) INK4d-deficient mice are fertile despite testicular atrophy. Moll Cell Biol 20: 372–378
66. Schulze-Gahmen U, Brandsen J, Jones HD, Morgan DO, Meijer L, Vesely J, Kim SH (1995) Multiple modes of ligand recognition: crystal structures of cyclin-dependent protein kinase 2 in complex with ATP and two inhibitors, olomoucine and isopentenyladenine. Proteins: Structure Function & Genetics 22: 378–91
67. de Azevedo WF, Mueller-Dieckmann H-J, Schultze-Gahmen U, Worland PJ, Sausville E, Kim S-H (1996) Structural basis for specificity and potency of a flavonoid inhibitor of human CDK2, a cell cycle kinase. Proc Natl Acad Sci USA 93: 2735–40
68. Lawrie AM, Noble ME, Tunnah P, Brown NR, Johnson LN, Endicott JA (1997) Protein kinase inhibition by staurosporine revealed in details of the molecular interaction with CDK2. Nature Struct Biol 4: 796–800

5 Genetic Manipulation and Transplantation of Male Germ Cells

T. Guillaudeux, C. Celebi, P. Auvray, B. Jégou

5.1 Introduction 59
5.2 Mature Spermatozoa As a New Vector for Transgenesis 60
5.3 Germ Cell Transplantation After a Possible *In Vitro* Transfection of Recombinant DNA 63
5.4 Microinjection of Recombinant DNA *In Situ* 63
5.5 Experiments in Progress in Our Laboratory 64
5.6 Conclusion 66
References 66

5.1 Introduction

In order to generate transgenic animals, a commonly used technique is the microinjection of linearized recombinant DNA into the male pronucleus of fertilized eggs. Derived embryos are then implanted into the oviduct of pseudopregnant females [1]. However, other methods have been developed in different laboratories during the last two decades. Some have used retroviral approaches, while others have taken advantage of *in vitro* manipulations of embryonic stem cells. More recently, some intensive work has been pursued in the field of animal cloning.

An alternative to these classical approaches is the genetic engineering of sperm. Thus, these germ cells, treated in various ways could then become usable as vectors for transgenesis. There are, at least, two reasons which render this perspective attractive: first, the "classical" methods of transgenesis mentioned above are very well adapted to the mouse, but they are often difficult to apply to other animals [2] and second, the approaches using sperm if proved successful, will be less traumatic and an easier way to introduce foreign DNA to an organism, especially in mammals.

Three approaches have been developed by different laboratories in order to generate transgenic animals using sperm as a vector:

1. mature spermatozoa, treated or not with different procedures to destabilize the plasma membrane, are mixed *in vitro* with exogenous DNA prior to natural or artificial fertilization.
2. male germ cells are isolated from a donor animal and then cultured *in vitro*. During this culture phase, cells may be transfected with exogenous DNA before their implantation in the seminiferous tubules of a recipient animal. The aim here is to obtain, after a complete spermatogenetic cycle, some mature spermatozoa derived from the reimplanted transfected germ cells and containing the exogenous DNA which would naturally or artificially be able to fertilize oocytes.
3. the recombinant DNA is directly microinjected into the seminiferous tubules of an animal, with the objective to facilitate the integration of the transgene into the genome of some stem cells *in vivo*. After a complete spermatogenetic cycle, it would be hoped that some spermatozoa carrying the exogenous DNA would be viable to fertilize oocytes.

5.2 Mature Spermatozoa As a New Vector for Transgenesis

The first attempt to produce transgenic animals using mature sperm was performed in 1971 on rabbits [3]. Since then, a number of groups have applied this technique mostly on mice, with varying degrees of success. Table 1 is an historical summary of work involving the genetic manipulation of mature spermatozoa.

Most of the early work carried out with mature spermatozoa was done using mice. In 1989, some doubt was cast on the validity of the methodology described by Lavitrano *et al.* [4], when Brinster *et al.* [5] were unable to repeat the earlier work. In the following decade, other

Table 1. Genetic manipulation of mature spermatozoa

Date	Discovery	Technology	Reference
1971	DNA incorporation in rabbit sperm heads	Incubation of sperm with exogenous DNA	[3]
1989	Incorporation of exogenous DNA in mouse sperm	Incubation of sperm with exogenous DNA	[4]
1989	Previous work [4] not reproduced	Same approach as [4] on a larger set of animals	[5]
1991	Incorporation of exogenous DNA in mouse spermatozoa	Liposome-mediated DNA uptake by spermatozoa	[6]
1998	Incorporation of exogenous DNA into mouse sperm (same authors and approach as in [4])	Incubation of sperm with exogenous DNA	[7]
1998	Injection of dead spermatozoa in mouse oocytes	ICSI (Intra Cytoplasmic Sperm Injection)	[8]
1999	Incorporation of exogenous DNA in mouse progeny	Incubation of freeze/dried, freeze/thawed or detergent treated sperm with exogenous DNA prior to ICSI	[9]
2000	Incorporation of exogenous DNA in rhesus macaque embryos	Incubation of sperm with exogenous DNA prior to ICSI	[10]

investigations were pursued using other animal models. Table 2 summaries the different species in which the sperm transfection methodology was tested. An estimate of the different efficiency rates for the production of transgenic animals is also presented. In these different studies, important variabilities in the ability of spermatozoa to incorporate and then transmit exogenous DNA, occur between species. However, in some cases work done on species like xenopus [16], abalone [17] and more recently mouse [9] showed that the use of mature sperm as a vector for transgenesis could be very efficient. The mouse work, with sperm treated with a detergent or submitted to several cycles of freezing and thawing or freezing and drying previous to incubation with DNA, led to the production of transgenic mice [9]. Two to 3% of oocytes microinjected with DNA treated sperm (ICSI) gave rise to transgenic founders

Table 2. Sperm transfection with exogenous DNA in different species

Species	Technique	Results	Reference
Rabbit	Incubation of DNA with spermatozoa	– Incorporation in spermatozoa heads, – Offspring = not tested	[3]
Boar and Bull	Incubation of DNA with spermatozoa	– Transgene detected in 39–78% of spermatoza in the post-acrosomal region, – Offspring = not tested	[11]
Pig	Incubation of DNA with spermatozoa	– 12% of the offspring is transgenic but with mosaïcism → no transgene found in the following generation	[12]
Pig	Electrotransfert of DNA in spermatozoa	– 75% of spermatozoa incorporate the transgene in the post-acrosomal region	[13]
Rainbow trout	Incubation of DNA with spermatozoa	Unsuccessful	[14]
Common carp African Catfish tilapia	Electrotransfert of DNA in spermatozoa	– 3–4% of the offspring are transgenic	[15]
Xenopus	Spermatozoon's nucleus +Restriction enzyme +DNA+ICSI	– 36% of the offspring are transgenic	[16]
Abalone	Electrotransfert of DNA in spermatozoa	– 65% of the offspring are transgenic	[17]

which were then able to transmit the exogenous DNA in a Mendelian fashion. Perry *et al.* [9] clearly showed that they were able to generate for the first time, using the spermatozoon approach, transgenic mammals which incorporated exogenous DNA inside their genome. As a consequence of this significant advance in genetic manipulation, many laboratories around the world are aiming to acquire this technology. For example, embryos of rhesus macaques expressing a transgene, have recently been produced [10].

5.3 Germ Cell Transplantation After a Possible *In Vitro* Transfection of Recombinant DNA

This second approach, using the male gamete as a natural vector for transgene insertion, was mostly developed in Ralph Brinster's laboratory. While the technology is a little more tedious to master than the first approach, it offers new technological perspectives. The basis of this approach, which is described in details in the following chapter of this book by Schlatt and collaborators, is the microinjection of germ cells obtained from a donor to a sterile recipient animal. Brinster and colleagues were able to obtain the establishment of spermatogenesis originating from the introduced germ cells, in 18–36% of the microinjected mice [18,19]. While this technology might one day, be used to treat severe sterility in man, a more immediate application of this discovery is likely to be the generation of transgenic animals. As a matter of fact, one can envisage to be able to transfect exogenous DNA in a male germ cell preparation obtained from a donor and then microinject only cells that have integrated the transgene in their genome. Recently, Brinster's team was able to establish germ cell cultures *in vitro*, for more than three months, using a co-culture system with feeder cells [20]. This is a necessary prerequisite for the generation of transgenic animals using this technology.

Starting from these significant advances, it seems likely that in a near future transgenic animals and in particular mice will be generated using the same type of *in vitro* genetic manipulation of male germ cells.

5.4 Microinjection of Recombinant DNA *In Situ*

In contrast to the previous approaches, this third method which is performed *in vivo*, involves the direct introduction of foreign DNA into germ stem cells, in the seminiferous tubules, using a microinjection procedure employing fine pipettes containing exogenous DNA mixed with different carriers. This approach allows spermatogenic stem cells carrying exogenous DNA in their genome to be obtained. These modified cells would then transmit the transgene to mature spermatozoa and hopefully to following generations. Kim *et al.* (1994) [21] showed that they were able to introduce exogenous DNA in male germ cells, using this method.

Fig. 1a–c. Paraffin embedded testis sections after microinjections of pCH110 + liposome X into mice seminiferous tubules. The β-galactosidase activity is revealed with standard procedures [22]. **a** Control. **b** 48 hours after injection. **c** 40 days after injection

5.5 Experiments in Progress in Our Laboratory

We are currently pursuing investigations using the approach described just above (see chapter 5.4). We decided to focus our first efforts on the *in vivo* microinjection technique, since we can reasonably expect to be able to generate some transgenic animals this way. A second rational for us to develop this technology was the possibility of transiently expressing a transgene in testicular cells: a long term goal is a transitory rescuing of fertility in males with genetic defects, which are affecting either Sertoli or germ cell function. In Table 3, we summarize the different transfecting products we experimented to facilitate the incorporation of our transgene into testicular cells. The vector we used was the circular plasmid expression vector pCH110 containing the *lac*Z reporter gene. We also tested an adenovirus vector in order to estimate its transitory expression efficiency, when injected in seminiferous tubules. Several mice were microinjected with different mixtures of plasmid and transfecting agents (liposomes). Appropriate sections were then tested for β-galactosidase activity: 48–96 h, 25 and 40 days after microinjection (Fig. 1). The optimum conditions were selected for routine use and so obtained mice were then mated with wild type females. The offspring were then tested by PCR using a specific primer pair for the *lac*Z gene and by Southern blotting to detect genomic integration of the transgene. Positive progeny, identified by PCR, were then mated with wild type mice. The new generation obtained was also tested for transgene integration. It is clear from the data obtained so far (Table 3) that we have been able to generate mice with transient transmission of the *lac*Z carrying vector using the *in situ* microinjection procedure.

Table 3. In situ microinjection of exogenous DNA in mouse seminiferous tubules

Mode of Transfection	Vector	β-galactosidase activity			Mating	F 0 Generation		F1 Generation	
		48/96 h	24 d	40 d		PCR	Southern Blot	PCR	Southern Blot
Electroporation	pCH110	(–)	(–)						
Retrovirus	AdCMV *lac Z*-cyto	(+++)	(++)						
PEI (Sigma)	pCH110	(±)							
Cellfectine (Gibco)	pCH110	(–)							
Lipofectine (Gibco)	pCH110	(–)							
Lipofectamine (Gibco)	pCH110	(±)		(+)					
DMRIE-C (Gibco)	pCH110	(–)							
A (our laboratory)	pCH110	(++)	(++)	(++)	OK	in progress	in progress	in progress	in progress
B (our laboratory)	pCH110	(±)							
X (our laboratory)	pCH110	(+++)	(++)	(++)	OK	(+)	(+)	in progress	in progress
Y (our laboratory)	pCH110	(+++)	(++)	(++)	OK	(+)	(+)	in progress	in progress
Z (our laboratory)	pCH110	(±)							

AdCMV *lacZ*-cyto: Adenovirus with a CMV promotor and containg the *lac Z* reporter gene, this vector was kindly provided by the Centre Hospitalier Universitaire-Laboratoire de Thérapie génique Bâtiment Jean Monnet-Hôtel Dieu, 30 Boulevard Jean Monnet, 44035 Nantes, PEI Cellfectine, Lipofectine, Lipofectamine and DMRIE-C are commercial liposomes; A and B are new liposomes, not commercialized and produced by the CNRS – Unité mixte de recherche no. 6521 – Faculté des Sciences et Techniques, 6 Avenue Le Gorgeu, BP 809, 29285 Brest Cedex; X, Y and Z are produced by the Chemistry Laboratory, ENSCR, 35000 Rennes. The β-galactosidase intensity is estimated from unrevealed (–), to intermediate (+), high (++) and very high (+++)

5.6 Conclusion

The new field of male germ cell transgenesis, even if some of the data are still at a preliminary stage, seems very promising, not only for research but also for animal husbandry.

The most advanced results come from experiments performed in mice with mature sperm using ICSI, the mastering of which seems relatively easy to acquire. At present, the success rate is no better than can be expected from traditional pronucleus microinjection, although we can hope that future improvement of this nascent technology will significantly increase its efficiency.

Since it is now clear that transgenic embryos can be produced by the method using DNA incubation with mature spermatozoa followed by ICSI (Table 1), it is essential that contamination by exogenous DNA be prevented at all costs. This caveat may be especially relevant to the work of IVF/Fertility clinics.

Acknowledgements. We are grateful to Dr D. Dresser, GERM-INSERM U435, Rennes for his precious comments on this article. We would like also to thanks the Laboratory of chemistry CNRS – Unité mixte de recherche n°6521-Faculté des Sciences et Techniques, 6 Avenue Le Gorgeu – BP 809–29285 Brest Cedex; and the Laboratory of Chemistry ENSCR, 35000 Rennes which kindly provided the non commercial liposomes. This work was supported by INSERM, Région Bretagne, Fondation pour la Recherche Médicale and Ministère de l'Education, de la Recherche et de la Technologie.

References

1. Gordon JW, Ruddle FH (1983) Gene transfer into mouse embryos: production of transgenic mice by pronuclear injection. Methods Enzymol. 101:411–433
2. Wall RJ, Kerr DE, Bondioloi KR (1997) Transgenic dairy cattle : genetic engeneering on a large scale. J. Dairy Sci 80:2213–2224
3. Brackett BG, Baranska W, Sawicki W, Koprowski H (1971) ^{3}H Uptake of heterologous genome by mammalian spermatozoa and its transfer to ova through fertilization. Proc Natl Acad Sci USA 68:353–357

4. Lavitrano M, Camaioni A, Fazio VM, Dolci S, Farace MG, Spadafora C (1989) Sperm cells as vectors for introducing foreign DNA into eggs: genetic transformation of mice. Cell 57:717–723
5. Brinster RL, Sandgren EP, Behringer RR, Palmiter RD (1989) No simple solution for making transgenic mice. Cell 59:239–241
6. Bachiller D, Schellander K, Peli J, Ruther U (1991) Liposome-mediated DNA uptake by sperm cells. Mol Reprod Dev 30:194–200
7. Maione B, Lavitrano M, Spadafora C, Kiessling AA (1998) Sperm-mediated gene transfer in mice. Mol Reprod Dev 50:406–409
8. Wakayama T, Yanagimachi R (1998) Development of normal mice from oocytes injected with freeze-dried spermatozoa. Nat Biotechnol 16: 639–641
9. Perry AC, Wakayama T, Kishikawa H, Kasai T, Okabe M, Toyoda Y, Yanagimachi R (1999) Mammalian transgenesis by intracytoplasmic sperm injection. Science 284:1180–1183
10. Chan AW, Luetjens CM, Dominko T, Ramalho-Santos J, Simerly CR, Hewitson L, Schatten G (2000) Foreign DNA transmission by ICSI: injection of spermatozoa bound with exogenous DNA results in embryonic GFP expression and live rhesus monkey births. Mol Hum Reprod 6:26–33
11. Camaioni A, Russo MA, Odorisio T, Gandolfi F, Fazio VM, Siracusa G (1992) Uptake of exogenous DNA by mammalian spermatozoa: specific localization of DNA on sperm heads. J Reprod Fertil 96:203–212
12. Lauria A, Gandolfi F (1993) Recent advances in sperm cell mediated gene transfer. Mol Reprod Dev 36:255–257
13. Horan R, Powell R, Bird JM, Gannon F, Houghton JA (1992) Effects of electropermeabilization on the association of foreign DNA with pig sperm. Arch Androl 28:105–114
14. Chourrout D, Perrot E (1992) No transgenic rainbow trout produced with sperm incubated with linear DNA. Mol Mar Biol Biotechnol 1:282–285
15. Muller F, Ivics Z, Erdelyi F, Papp T, Varadi L, Horvath L, Maclean N (1992) Introducing foreign genes into fish eggs with electroporated sperm as a carrier. Mol Mar Biol Biotechnol 1:276–281
16. Kroll KL, Amaya E (1996) Transgenic *Xenopus* embryos from sperm nuclear transplantations reveal FGF signaling requirements during gastrulation. Development 122:3173–3183
17. Tsai HJ, Lai CH, Yang HS (1997) Sperm as a carrier to introduce an exogenous DNA fragment into the oocyte of Japanese abalone (*Haliotis divorsicolor suportexta*). Transgenic Res 6:85–95
18. Brinster RL, Zimmermann JW (1994) Spermatogenesis following male germ-cell transplantation. Proc Natl Acad Sci USA 91:11298–11302

19. Brinster RL, Avarbock MR (1994) Germline transmission of donor haplotype following spermatogonial transplantation. Proc Natl Acad Sci USA 91:11303–11307
20. Nagano M, Avarbock MR, Leonida EB, Brinster CJ, Brinster RL (1998) Culture of mouse spermatogonial stem cells. Tissue & Cell 30:389–397
21. Kim JH, Jung-Ha HS, Lee HT, Chung KS (1997) Development of a positive method for male stem cell-mediated gene transfer in mouse and pig. Mol Reprod Dev 46:515–26
22. Mansour S, Thomas KR, Deng C, Capecchi MR (1990) Introduction of a lacZ reporter gene into the mouse int-2 locus by homologous recombination. Proc Natl Acad Sci USA 87:7688–7692

6 Germ Cell Culture, Genetic Manipulation and Transplantation

S. Schlatt, A. G. Schepers, V. von Schönfeldt

6.1 Introduction 69
6.2 Germ Cell Culture and Genetic Manipulation 71
6.3 Germ Cell Transplantation As a Research Tool 78
6.4 Germ Cell Transplantation As a Clinical Tool 81
6.5 Conclusion 82
References 83

6.1 Introduction

Only the male, but not the female adult gonad contains mitotically active germ cells. The presence of proliferatively active spermatogonial stem cells offers a potential for genetic engineering of the male germ line. Despite of stem cells, the active seminiferous epithelium contains several layers of different germ cell types. Their classification is depending on the developmental stage: spermatogonia represent the diploid stem cells and the premeiotic differentiating germ cells. All cells in prophase and divisons of meiosis are classified as spermatocytes. Postmeiotic germ cells are round or elongating spermatids. The process of spermatogenesis is a genetically programmed and timewise well defined series of proliferation and differentiation steps. The fate of all germ cells which have started this process is either differentiation into mature spermatids or apoptotic cell death.

For transfection of cells and more important for the selection of transfected cells, it is mandatory to achieve high proliferation rates of the target cells. Only the true spermatogonial stem cells, but not the differentiating spermatogonia, spermatocytes or spermatids can undergo unlimited rounds of mitotic divisions.

Therefore, approaches for stable transfection and subsequent selection of clones through the male germ line are targeting the spermatogonial stem cells or differentiating spermatogonia during the period of premeiotic germ cell expansion. Other approaches of genetic engineering by injection of DNA into seminiferous tubules would not allow to select stably transfected clones of cells, although they might allow to introduce a transgene to the germ line (Yamazaki et al., 1998). This review focusses on possible future pathways for the stable transfection of the male germ cell line and therefore puts spermatogonia in the centre of interest.

Although spermatogonia have regained sientific interest over the last few years, our knowledge on spermatogonial stem cells is still limited. While type A-spermatogonia are morphologically well chracterised for several rodent and primate species, the knowledge of their function and physiology is very limited mainly due to the lack of good approaches for enrichment and culture of A-spermatogonia (for review: Meistrich and van Beek, 1993, deRooij 1998; deRooij and Grootegoed, 1998).

This review focusses on the late methodological breakthroughs in the transplantation of spermatogonia which have led to the revival of research on testicular stem cells. Furthermore this review highlights some problems that remain to be solved before testicular stem cells are used to introduce transgenes into the male germ line facilitating the production of transgenic life stock. Germ cell transplantation might also become a tool for clinicians offering an extracorporal reserve of stem cells to restore fertility after gonadotoxic chemo-or radiotherapeutic treatment of oncological patients.

6.2 Germ Cell Culture and Genetic Manipulation

6.2.1 Isolation of Spermatogonia

Different subclasses of spermatogonia populate the basement membrane of the seminiferous epithelium. While most of these cells are differentiating spermatogonia which have entered the process of spermatogenesis and pass through a species-dependent series of mitotic divisions and differentiation steps, only few undifferentiated A-spermatogonia remain as true stem cells undergoing self-renewing divisions. Two models (A_S-model: Huckins, 1971; clone fragmentation model: Erickson, 1981) exist for stem cell renewal and spermatogonial differentiation (for review: Meistrich and van Beek, 1993). The study of the physiology and roles of the various subtypes is hampered by the enrichment of mixed populations where the true stem cells might present a small minority of the isolated spermatogonia.

Specific cellular markers and strategies to differentiate between the various spermatogonial subtypes do not yet exist. However, in vitro experiments on isolated spermatogonia depend on efficient approaches to enrich and culture the different subtypes of spermatogonia. So far, this has not been possible. Testes from immature animals have been used in the past to isolate spermatogonia. Isolation procedures using elutriation (Bucci et al., 1986) or velocity sedimentation in a BSA-gradient at unit gravity (Bellvé et al., 1971) were the conventional techniques for spermatogonial isolation leading to the above mentioned problem of enriching mixed populations of spermatogonial cells. Isolation and culture of neonatal gonocytes have been described (Orth and McGuiness, 1991). In vitro, no further progression and differentiation into A-spermatogonia occurred and thus did not allow to study several aspects of A-spermatogonial physiology.

Magnetic cell separation has become a widespread tool for the enrichment of different cell types from the blood cell line and other organs. We have applied this technique in order to isolate spermatogonia from the testis and used c-kit antibodies as a marker. An enrichment of spermatogonia up to 25–54% was achieved even when fully mature testes from Djungarian hamsters, mice and marmoset monkeys were used (von Schönfeldt et al., 1999). Since the enrichment efficiency was similar in regressed and non-regressed testes, this isolation strategy appeared to be independent of the number of non-target cells in the digested cell preparation. C-kit antibodies were chosen as the marker for the cell sorting since several observations support the presence of the receptor for stem cell factor on spermatogonia (Yoshinaga et al., 1991; Dym et al, 1995; Brinster and Avarbock, 1994). Furthermore, polyclonal rabbit anti c-kit antibodies are commercially available. Using a c-kit antibody and magnetic separation columns enabled us to label and sort fractions of testicular cells with a recovery of around 10^5 cells in the magnetic fraction per 10^7 cells sorted. Depending upon the preparation up to 55% of these cells were recognised as c-kit positive by flow cytometry. The viability of the cells ranged from 82% to 91%. The efficiency of our spermatogonial separation procedure was slightly lower in comparison to the elutriation procedure described for immature rats (Bucci et al., 1986). However, the superiority of magnetic cell sorting was obvious when normal testes showing the full germ cell complement were used for the separation procedure. Why was it mandatory to use testes from immature animals for both centrifugal elutriation and sedimentation velocity approaches. Since the differences in size and shape of the cells determine the isolation efficiency, spermatocytes and round spermatids are quite similar in comparison to spermatogonia which therefore makes it impossible to select and isolate the comparatively small number of spermatogonial cells from fully mature testes.

It is important to note that all subtypes of spermatogonia form small clones of cells which do not separate completely after mitotic divisions and remain in contact by cytoplasmic bridges. Therefore, the majority of spermatogonia will be eluted in small aggregates. This limits drastically the isolation efficiency using elutriation. The presence of small fragments, however, does not interfere with the magnetic enrichment procedure highlighting another advantage of magnetic cell sorting for spermatogonia.

However, an important limitation of this immunological approach is the absence of specific antibodies for the various subtypes of spermatogonia. It was recently shown that c-kit is expressed in all differentiating spermatogonia but may be absent or expressed only at very low levels in undifferentiated spermatogonia (Schrans-Stassen et al., 1999). Therefore it is most likely that the use of c-kit antibodies as markers for magnetic cells separation leads to enrichment of differentiating spermatogonia. As the repopulation index decreases after the use of c-kit antibodies coupled to magnetic beads for isolation, it appears that c-kit is not immunologically detectable in true stem cells (Shinohara et al., 1999). There is urgent need for additional specific antibodies which can be used for the selection of testicular stem cells and the subtypes of differentiating spermatogonia.

6.2.2 Culture of Spermatogonia

Survival of spermatogonia under in vitro culture conditions is limited. Freshly isolated A-spermatogonia from 80 day old pigs died quickly when cultured in Dulbecco´s modified Eagle´s medium (Dirami et al., 1999). Increased survival rates (30–40% after 120 hours) were achieved with spermatogonia maintained in KSOM medium. The addition of soluble stem cell factor and more potently GMCSF improved survival rates indicating that SCF and GMCSF may act as survival factors for spermatogonia. Coculture of spermatogonia with Sertoli cells was often used to improve survival rates of spermatogenic cells in vitro (Tres and Kierszenbaum, 1983). Coculture experiments indicated that stem cell factor influences the proliferation of spermatogonia (Rossi et al., 1993). Whatever design is used for culturing spermatogonia, the major problem remaining is the heterogeneity of the cultured spermatogonia, the immense rate of spermatogonial death and the difficulty to maintain a high proliferation rate of these cells.

One reason why spermatogonia cannot be cultured for long periods became obvious when spermatogonia were observed in mixed cultures of digested seminiferous tubules. A coculture of Sertoli cells, peritubular cells and spermatogonia undergoes well defined morphogenetic changes during the first few days of culture (Schlatt et al., 1996). Spermatogonia are observed as small clones of cells on top of the other two cell types which form a mixed confluent monolayer (Fig. 1 a-f). Spermatogonia can be identified morphologically as small clones of two, four, eight or sixteen cells or immunohistochemically by their positive staining for c-kit (Fig. 1 b). All cells inside such clones show complete synchronism as depicted by their identical chromatin pattern and simultaneous mitotic divisions (Fig. 1 e,f). Although progression into larger clones can be observed, no entry into meiosis occurs and many of the clones undergo apoptosis. Slowly but steadily with time, these cultures are

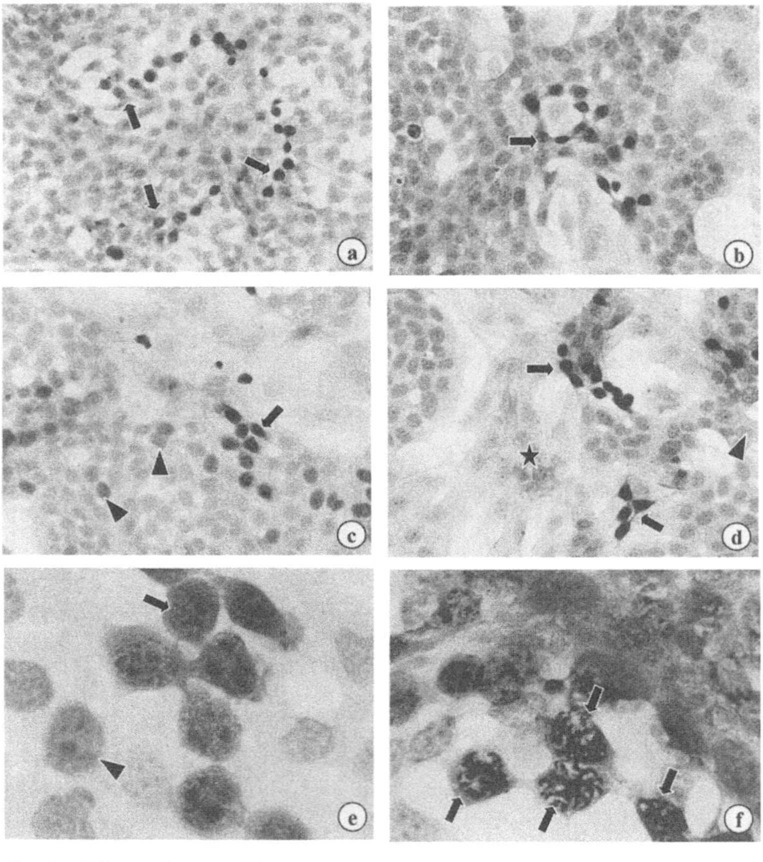

Fig. 1a–f. Legend see p. 75

depleted of spermatogonia, since no new clones of cells develop spontaneously while the more mature clones are lost through apoptotic cell death. These changes can be observed under very basic culture conditions in Dulbecco´s modified Eagle´s medium with no serum, growth factors or hormones present. This indicates that the developmental program of differentiating spermatogonia, once initiated, is maintained and lasts as long as survival and progression of the cells are favoured to the initiation of apoptosis. However, these culture conditions do not provide the stimuli for the differentiation of A1-spermatogonia from undifferentiated stem cell spermatogonia or the initiation of the meiotic prophase (Fig. 2).

So far, no approach using fully digested spermatogonia has been presented to replenish spermatogonial cells under in vitro conditions or to initiate the meiotic process. The lack of such a system has important implications on the ability to transfect spermatogonia. The mitotically inactive stem cells can not be easily transfected while the rapid proliferation of differentiating spermatogonia might not leave enough time for transfection since the process terminates always with apoptotic cell death. It will be important to improve the culture system for spermatogonia to widen the window for the manipulation of these cells either by restoring stem cell proliferation and expansion or by inducing meiosis in

Fig. 1a–f. Micrograph showing cocultures of Sertoli cells (*arrowheads*), peritubular cells (*star*) and spermatogonia after 4–9 days of culture in DMEM. **a, c–f** Immunohistochemically stained for alpha smooth muscle actin as a marker for peritubular cells and bromodeoxyuridine as a marker of cells in S-phase of the cell cycle. **b** Immunohistochemically stained for c-kit. **a** After 4 days of culture, small clones of spermatogonia interconnected by cytoplasmic bridges (*arrows*) are recognised by their distinctive and identical nuclear morphology on top of a mixed confluent layer of Sertoli cells and peritubular cells. **b** Positive immunohistochemical staining of c-kit in spermatogonia (*arrow*) after 5 days of culture. **c, d** BrdU-labelling and immunohistochemical nuclear staining reveals proliferative activity of Sertoli cells (*arrowheads*) and clones of spermatogonia (arrow) after 8–9 days of culture. Peritubular cells are recognised by their labelled bundles of actin (star). **e** A high power magnification of c showing the clone of 8 BrdU-positive A4 spermatogonia (*arrow*). The cytoplasmic bridges as well as the identical nuclear chromatin pattern of differentiating spermatogonia are visible. A BrdU-positive Sertoli cell in a different plain of focus is indicated (*arrowhead*). **f** The complete synchronism of cell cycles in interconnected spermatogonial clones (arrow) is seen during mitosis

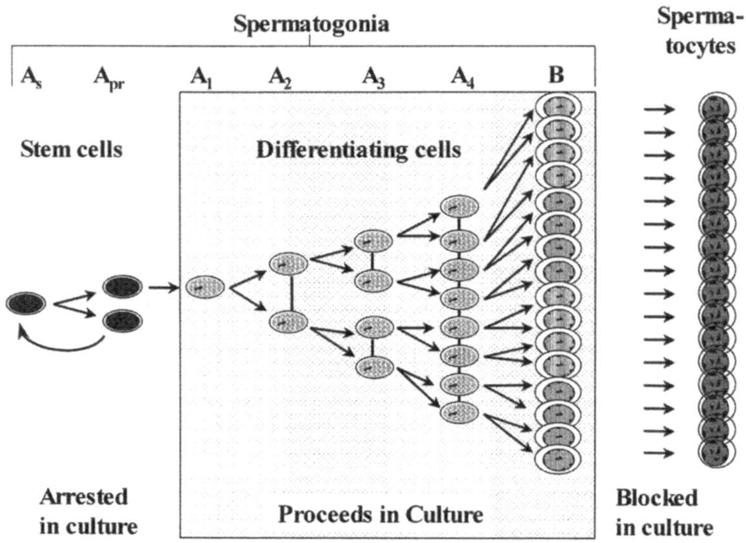

Fig. 2. Schematic representation of premeiotic germ cell development in the rat The kinetics and labelling of spermatogonia are taken from the As-model. (Huckins, 1971). The shaded area indicates phases of premeiotic expansion which occur spontaneously in culture without addition of serum or growth factors. The differentiation of new A1 spermatogonia as well as entry into meiosis does not occur under these conditions

vitro (compare Fig. 2). It was shown that successful germ cell transplantation is possible after culture of mouse spermatogonial stem cells in coculture with feeder cells for more than four months (Nagano et al., 1998). This indicates that small numbers of true stem cells survive long periods in culture. Supposedly, these cells were mitotically silent until after the transplantation. Transplantation of spermatogonia, which will be discussed later, presents an alternative to in vitro meiosis for the production of mature gametes from spermatogonial stem cells.

To overcome the problem of acute loss of spermatogonia during the isolation and early culture period it is possible to culture small testicular fragments or seminiferous tubules. Organ culture was developed many years ago to study various aspects of testicular physiology (Steinberger et al, 1964a, b). In one of these studies, it was observed that testicular tissue fragments grown in defined medium retained the tubular architecture for up to 8 months. In these cases the structural elements of the wall of the seminiferous tubules, the Sertoli cells, and their topography were maintained (Steinberger et al., 1970). More recently, an organ culture system has been shown to support germ cell proliferation and differentiation in short term culture (Boitani et al, 1993), and Sertoli cell proliferation was observed to exhibit developmentally regulated changes in response to the addition of hormones and growth factors (Boitani et al., 1995). We have performed a study analysing and comparing the first ten days of rat testicular development which includes the period of premeiotic germ cell expansion (Schlatt et al., 1999). It was shown that spermatogonia continued to proliferate under very basic conditions over a three day culture period. Recent studies support the use of organ culture for the analysis of spermatogonial behaviour showing that stem cell factor prevents germ cell apoptosis and stimulates spermatogonial DNA synthesis (Hakovirta et al., 1999; Yan et al., 2000). It will be interesting to further analyse the effects of other hormones or growth factors like FSH or activin in such an organ culture system. The use of this system for the transfection of the male germ line has not been tested.

In conclusion, strategies for genetic manipulation of spermatogonia will depend on the development of new culture and transfection systems. The isolation and culture techniques for spermatogonia do not fullfill the requirements for a useful and efficient approach to transfect the male germ line. There was a surprisingly fast development of new techniques that facilitated the transplantation of spermatogonia and thereby achieve progression of transplanted cells from stem cells to fertile sperm. The future development of appropriate isolation and culture techniques for spermatogonia opens new pathways to a possible transfection of the male germ line.

6.3 Germ Cell Transplantation As a Research Tool

6.3.1 The Development of the Technique

The pioneer work for the development of germ cell transplantation was performed by Brinster and Zimmermann (1994) and Brinster and Avarbock (1994). Donor spermatogenesis, recognised by developing germ cells carrying the lac-Z gene, was restored from spermatogonial stem cells microinjected into seminiferous tubules of host animals (Brinster and Zimmermann, 1994). Offspring showing the donor haplotype was generated in mating experiments (Brinster and Avarbock, 1994; Ogawa et al., 2000). Electron microscopy of the host testes showed focal arrangements of quantitatively and qualitatively normal donor spermatogenesis (Russell et al., 1996). A time course study after transfer showed that most transplanted cells degenerate and disappear from the testis during the first few weeks (Parreira et al., 1998). The first meiotic donor germ cells appeared after one month and two months after transplantation only 1% of all seminiferous tubules contained meiotic germ cells. Xenogeneic transfer of rat germ cells into mouse testis prompted rat germ cells to associate with mouse Sertoli cells (Clouthier et al., 1996). Mouse Sertoli cells were able to support the developing rat germ cells. Their differentiation occurred according to the kinetics and topography typical for the rat (Franca et al., 1998). Transplanted spermatogonia from hamster, rabbits or dogs were not able to fully restore spermatogenesis in the immunodeficient mouse testis (Ogawa et al., 1999a; Dobrinski et al., 1999). The hamster to mouse transfer resulted in the production of abnormal hamster sperm. Although rabbit or dog spermatogonia repopulated the basal compartment of the host seminiferous tubules, no meiotic germ cells of either species were produced. Although it appears more difficult to prepare rats as hosts for the transplantation of mouse germ cells than vice versa, it was shown that mouse germ cells populate the rat testis and that donor-derived mouse spermatogenesis is initiated in the rat testis (Ogawa et al., 1999b). A summary of the most important publications on germ cell transplantation is presented in Table 1.

Table 1. Development of germ cell transplantation

Year	Autor	Title of publication
1994	Brinster and Avarbock	Germline transmission of donor haplotype in mice.
1994	Brinster and Zimmermann	Spermatogenesis following male germ cell transplantation.
1995	Jiang and Short	Male germ cell transplanation in rats: apparent synchronisation of spermatogenesis between host and donor seminiferous epithelia.
1996	Clouthier et al.	Rat spermatogenesis in mouse testis
1996	Avarbock et al.	Reconstitution of spermatogenesis from frozen spermatogonial stem cells.
1998	Franca et al	Germ cell genotype controls cell cycle during spermatogenesis in the rat.
1998	Parreira et al.	Development of germ cell transplants in mice.
1998	Nagano et al.	Culture of mouse spermatogonial stem cells.
1999	Schlatt et al.	Germ cell transfer into rat, bovine, monkey and human testes.
1999	Nagano et al.	Pattern and kinetics of mouse donor spermatogonial stem cell colonization in recipient testes.
1999	Dobrinski et al.	Transplantation of germ cells of rabbits and dogs into mouse testis.
1999a	Ogawa et al.	Xenogeneic spermatogenesis following transplantation of hamster germ cells to mouse testes.
1999	Shinohara et al.	Beta1- and alpha6-integrin are surface markers on mouse spermatogonial stem cells.
1999b	Ogawa et al.	Recipient preparation is critical for spermatogonial transplantation in the rat
2000	Ogawa et al.	Transplantation of male germ line stem cells restores fertility in infertile mice.

6.3.2 New Findings Arising from Experiments Using Germ Cell Transplantation

From these experiments, it became obvious that the blood testis barrier of the adult testis did not hinder stem spermatogonia to migrate from the adluminal compartment to the basal compartment of the seminiferous epithelium. The appearance of sperm in w/w mutant mice arising from transplanted spermatogonia revealed that germ cells without functional c-kit receptors are not able to enter the process of spermatogenesis (Brinster and Zimmermann, 1994). The important role of the c-kit-stem cell factor system was substantiated by transplanation of spermatogonia from infertile mice carrying a mutation of the stem cell factor gene into infertile mice with a mutated c-kit receptor gene. The presence of Sertoli cells producing membrane bound stem cell factor and germ cells expressing functional c-kit receptors leads to qualitatively normal repopulation of the testes with developing germ cells (Ogawa et al., 2000). Although the number of tubules containing A-spermatogonia does not change in c-kit mutants three months after germ cell transplantation, more than 25% of the seminiferous tubules contain transplanted spermatogenic cells (Parreira et al., 1998).

Various approaches for the preparation and handling of the donor cell suspensions were described. As an alternative to intratubular infusion of germ cells by microinjection, injections into the efferent ducts as well as into the rete testis of the host mice have been succesfully employed (Ogawa et al., 1997). The ability to cryopreserve the donor germ cells for long periods prior to the transfer was established (Avarbock et al., 1996, Nagano and Brinster, 1998). The transplantation of spermatogonial stem cells cultured for four months was successful. As indicated earlier, this opens future applications for in vitro multiplication and manipulation of male germ line cells (Nagano et al., 1998).

6.4 Germ Cell Transplantation As a Clinical Tool

6.4.1 The need for New Gonadal Protection Strategies

Testicular cancer, leukemia and Hodgkin's disease are frequently observed in children and young adults of reproductive age. Improved diagnostic tools and rising success of surgery, radiation and chemotherapy over the past decades led to a considerable increase in the survival rate of these patients and a cure has become a realistic goal. Often, oncological therapies result in severe impairment of testicular function (Naysmith et al., 1998; Radford et al., 1999). The combined chemotherapy for Hodgkin's results in deterioration of testis function occurring in up to 90% of postpubertal males (Sherins and deVita, 1973; Roeser et al., 1978; Chapman et al., 1981; Waxman et al., 1982). Studies in mice showed that the rapidly dividing diploid spermatogonia are the most sensitive testicular cells to the cytotoxic effects of radiotherapy (Meistrich, 1993) and chemotherapy (Meistrich et al., 1982). In men, as yet, all hormonal attempts to protect the seminiferous epithelium by reducing the rate of spermatogenesis failed: no improvement in posttreatment fertility has been convincingly demonstrated (Meistrich et al., 1994; Johnson et al., 1985; Waxman et al., 1987). These data indicate the need for better fertility protection in these patients, as many of them would not be able to father a child at all and none of them would do so without assisted fertilisation techniques. Extracorporal storage of spermatogonial stem cells followed by germ cell transplantation might lead to a better recovery of spermatogenesis. Therefore, autologous germ cell transplantation might become a clinically important technique by which germ cells can be removed from and reintroduced into the male gonad

6.4.2 Preclinical Research into Germ Cell Transplantation

We established a technique for germ cell transfer in primate species using the cynomolgus monkeys (**Macaca fascicularis**) as model for the human testis.

Microninjection of seminiferous tubules could not be applied since the increase in the volume-to-surface ratio did not allow efficient filling of the testis and seminiferous tubules from calf, bull, monkey and man have a resistent lamina propria and a highly convoluted tubular mass (Schlatt et al., 1999). The easiest, least invasive and most efficient filling of the seminiferous tubules was obtained by injections into the rete testis of bull, monkey and man. The correct placement of the injection needle into the rete testis was controlled by ultrasonography. The backpressure of seminiferous tubule fluid in the active testis did not allow deep infusion of germ cells into the tubular compartment. However, good infiltration was achieved in involuted testes of monkeys which had been exposed to six weeks of GnRH-antagonist treatment (Schlatt et al., 1999). The immunohistochemical detection of B-spermatogonia four weeks after autologous germ cell transfer revealed the success of the germ cell transplantation (Schlatt et al., 1999). This transfer technique was also applicable to surgically removed human testes (Brook et al., 1998; Schlatt et al., 1999).

One major limitation to the application of germ cell transplantation in oncological patients is the safety aspect. In lymphoma and leukemia patients the testis is a likely organ for the settlement of metastasing cells. Reintroducing malignant cells into a patient who was previously cured of the disease must be excluded. In respect to this problem, magnetic cell separation techniques using antibodies for the detection of spermatogonia might become a powerful clinically used tool (Schönfeldt et al., 1999).

6.5 Conclusion

The combination of germ cell transplantation and in vitro cell culture techniques offer pathways for genetic engineering of the male germ line. However, better approaches for the characetrisation and labelling of the various spermatogonial subclasses, for the isolation and for the culture of spermatogonia are needed as only these techniques allow a wide range of new applications in the production of transgenic lifestock.

Before ethically acceptable guidelines are agreed on, a clinical application of germ cell transplantation has to be strictly detached from genetic manipulation of the transferred cells. However, while it may be deemed unethical to apply germ line manipulation in men, it can be argued that it would be also be unethical not to use germ cell transplantation as a tool to protect the fertility of oncological patients.

Acknowledgements. Our own work reported in this review was supported by several grants from the Deutsche Forschungsgemeinschaft and a joint grant from the German Academic Exchange Service (DAAD) and the British Council. The work was made possible by the tremendous support from Prof. Dr. E. Nieschlag at the Institute of Reproductive Medicine in Münster. Part of the work was initiated and performed with the help of Prof. Dr. D. deKretser and Dr. K.L. Loveland at the Institute of Reproduction and Development, Monash University, Clayton, Australia. We are indebted to Professor Dr. G.F. Weinbauer, Professor Dr. R. Gosden, Dr. C. Rolf, Dr. A. Kamischke, Dr. G. Rosiepen, Dr. P.F. Brook, Dr. L. Foppiani and I. Upmann, who contributed to the work presented.

References

Avarbock MR, BrinsterCJ, Brinster RL (1996) Reconstitution of spermatogensis from frozen spermatogonia stem cells. Nat Med 2: 693–696

Bellvé AR, Cavicchia JC, Millette O´Brien DA, Bhatnagar YM, Dym M (1977) Spermatogenic cell of the prepuberal mouse: isolation and morphological characterization. J Cell Biol 74: 68–85

Boitani C, Politi MG, Menna T (1993) Spermatogonial cell proliferation in organ culture of immature rat testis. Biol Reprod 48: 761–767

Boitani C, Stefanini M, Fragale A, Morena AR (1995) Activin stimulates Sertoli cell proliferation in a defined period of rat testis development. Endocrinology 136: 5438–5444

Brinster RL, Avarbock MR (1994) Germline transmission of donor haplotype following spermatogonial transplantation. Proc Natl Acad Sci USA 91: 11303–11307

Brinster RL, Zimmermann JW (1994) Spermatogenesis following male germcell transplantation. Proc Natl Acad Sci USA 91: 11289–11302

Brook PF, Schlatt S, Rosiepen G, Radford JA, Shallet SM, Gosden RG (1998) An infusion technique for germ cell transfer into the seminiferous tubules of the human testis. Society for the Study of Fertility, Glasgow, July 1998, J Reprod Fertil, abstract series 21, 45

Bucci LR, Brock WA, Johnson TS, Meistrich ML (1986) Isolation and biochemical studies of enriched populations of spermatogonia and early spermatocytes from rat testes. Biol Reprod 34: 195–206

Chapman RM, Sutcliffe SB, Malpas JS (1981) Male gonadal dysfunction in Hodgkins´s disease: a prospective study. JAMA 245, 1323–1328

Clouthier DE, Avarbock MR, Maika SD, Hammer RE, Brinster RL (1996) Rat spermatogenesis in mouse testis. Nature 381, 418–421

Dirami G, Ravindranath N, Pursel V, Dym M. (1999) Effects of stem cell factor and granulocyte macrophage-colony stimulating factor on survival of porcine type A spermatogonia cultured in KSOM. Biol Reprod 61: 225–230

Dobrinski I, Avarbock MR, Brinster RL (1999) Transplantation of germ cells from rabbits and dogs into mouse testes. Biol Reprod 61: 1331–1339

DeRooij DG (1998) Stem cells in the testis. Int J Exp Pathol 79: 67–80

De Rooij DG, Grootegoed JA (1998) Spermatogonial stem cells. Curr Opin Cell Biol 10: 694–701

Dym M, Jia M-C, Dirami G, Price JM, Rabin SJ, Mocchetti I, Ravindranath N (1995) Expression of the c-kit receptor and its autophosphorylation in immature rat type A-spermatogonia. Biol Reprod 52: 8–19

Franca LR, Ogawa T, Avarbock MR, Brinster RL, Russell LD (1998) Germ cell genotype controls cell cycle during spermatogenesis in the rat. Biol Reprod 59: 1371–1377

Huckins C (1971) The spermatogonial stem cell population in adult rats. I. Their morphology, proliferation and maturation. Anat Rec 169: 533–558

Hakovirta H, Yan W, Kaleva M, Zhang F, Vanttinen K, Morris PL, Soder M, Parvinen M, Toppari J (1999) Function of stem cell factor as a survival factor of spermatogonia and localisation of messenger ribonucleic acid in the rat seminiferous epithelium. Endocrinology 140: 1492–1498

Jiang F-X, Short RV (1995) Male germ cell transplantation in rats: apparent synchronisation of spermatogenesis between host and donor seminiferous epithelia. Int J Androl 18: 326–330

Johnson DH, Linde R, Hainsworth JD, Vale W, Rivier J, Stein R, Flexner J, van Welch R, Greco FA. (1985) Effect of a luteinizing hormone releasing hormone agonist given during combination chemotherapy on posttherapy fertility in male patients with lymphoma: preliminary observations. Blood 65: 832–836

Meistrich ML (1993) Effects of chemotherapy and radiotherapy on spermatogenesis. Eur Urol 23: 136–141

Meistrich ML, van Beek MEAB (1993) Spermatogonial stem cells. In: Desjardins C, Ewing LL (eds.) Cell and molecular biology of the testis, Oxford University Press, New York, pp 266–295

Meistrich ML, Finch M., da Cunha MF, Hacker U, Au WW (1982) Damaging effects of fourteen chemotherapeutic drugs on mouse testis Cancer Res 42: 122–131

Meistrich ML, Wilson G, Ye W-S, Kurdoglu B, Parchuri N, Terry N (1994) Hormonal protection from procarbazine-induced testicular damage is selective for survival and recovery of stem spermatogonia. Cancer Res 54: 1027–1034

Nagano M, Brinster RL (1998) Spermatogonial transplantation and reconstitution of donor cell spermatogenesis in recipient males. Acta Pathol Microsc Immunol Scand 106: 47–55

Nagano M, Avarbock MR, Leonida EB, Brinster C.J, Brinster, RL (1998) Culture of mouse spermatogonial stem cells. Tissue & Cell 30: 389–397.

Nagano M, Avarbock MR, Brinster RL (1999) Pattern and kinetics of mouse donor spermatogonial stem cell colonization in recipient testes. Biol Reprod 60: 1429–1436

Naysmith TE, Blake DA, Harvey VJ, Johnson NP (1998) Do men undergoing sterilizing cancer treatments have a fertile future? Hum Reprod 13: 3250–3255

Ogawa T, Arechaga JM, Avarbock MR, Brinster RL (1997) Transplantation of testis germinal cells into mouse seminiferous tubules. Int J Dev Biol 41: 111–122

Ogawa T, Dobrinski I, Avarbock MR, Brinster RL (1999a) Xenogeneic spermatogenesis following transplantation of hamster germ cells to mouse testes. Biol Reprod 60: 515–521

Ogawa T, Dobrinski I, Brinster RL (1999b) Recipient preparation is critical for spermatogonial transplantation in the rat. Tissue Cell 31: 461–472

Ogawa T, Dobrinski I, Avarbock, MR, Brinster RL (2000) Transplantation of male germ line stem cells restores fertility in infertile mice. Nat Med 6: 29–34

Parreira CG, Ogawa T, Avarbock MR, Franca LR, Brinster RL, Russell LD (1998) Development of germ cell transplants in mice. Biol Reprod 59: 1360–1370

Radford JA, Shalet SM, Lieberman BA (1999) Fertility after treatment for cancer. Brit Med J 319: 935–936

Rossi P, Dolci S, Albanesi C, Grimaldi P, Ricca R, Geremia R (1993) Follicle-stimulating hormone induction of steel factor (SLF) mRNA in mouse Sertoli cells and stimulation of DNA synthesis in spermatogonia by soluble SLF. Dev Biol 155: 68–74

Russell LD, Franca LR, Brinster RL (1996) Ultrastructural observations of spermatogenesis in mice resulting from transplantation of mouse spermatogonia. J Androl 17: 603–614

Schlatt S, deKretser DM, Loveland KL (1996) Discriminative analysis of rat Sertoli and peritubular cells and their proliferation: Evidence for follicle-stimulating hormone mediated contact inhibition of Sertoli cell mitosis. Biol Reprod 55: 227–235

Schlatt S, Rosiepen G, Weinbauer GF, Rolf C, Brook PF, Nieschlag E (1999) Germ cell transfer into rat, bovine, monkey and human testes. Human Reprod 14: 144–150

Schlatt S, Zhengwei Y, Meehan T, deKretser DM, Loveland KL (1999) Application of morphometric techniques to postnatal rat testes in organ culture: insights into testis growth. Cell Tiss Res 298: 335–343

Schönfeldt vV, Krishnamurthy H, Foppiani L, Schlatt S (1999) Magnetic cell sorting as a fast and efficient method of enriching viable spermatogonia from rodent and primate testes. Biol Reprod 61: 582–589

Schrans-Stassen BH, van de Kant HJ, de Rooij DG, van Pelt AM (1999) Differential expression of c-kit in mouse undifferentiated and differentiating type A spermatogonia. Endocrinology 140: 5894–5900

Sherins RJ, DeVita VT (1973) Effect of drug treatment for lymphoma on male reproductive capacity: studies of men in remission after therapy. Ann Intern Med 79: 216–220

Shinohara T, Avarbock MR, Brinster RL (1999) Beta1- and alpha 6-integrin are surface markers on mouse spermatogonial stem cells. Proc Natl Acad Sci USA 11: 5504–5509

Steinberger A, Steinberger E, Perloff WH (1964a) Mammalian testes in organ culture. Exp Cell Res 36: 19–27

Steinberger E, Steinberger A, Perloff WH (1964b) Studies on growth in organ culture of testicular tissue from rats of various ages. Anat Rec 148: 581–589

Steinberger E, Steinberger A, Ficher M (1970) Study of spermatogenesis and steroid metabolism in cultures of mammalian testes. Rec Progr Horm Res 26: 546–588

Tres LL, Kierszenbaum AL (1999) Cell death patterns of the rat spermatogonial cell progeny induced by Sertoli cell geometric changes and FAS (CD95) agonist. Dev Dyn 214: 361–371

Waxman JH, Terry YA, Wrigley PF, Malpas JS, Rees LH, Besser GM, Lister TA (1982) Gonadal function in Hodgkin's disease: long-term follow-up of chemotherapy. Brit Med J 285: 1612–1613

Waxman JH, Ahmed R, Smith D, Wrighley PFM, Gregory W, Shalet S, Crowther D, Rees LH, Besser GM, Malpas JS, Lister TA (1987) Failure to preserve fertility in patients with Hodgkin´s disease. Cancer Chemoth Pharmacol 19: 159–162

Yan W, Suominen J, Toppari J (2000) Stem cell factor protects germ cells from apoptosis in vitro. J Cell Sci 113: 161–168

Yamasaki Y, Fujimoto H, Ando H, Ohyama T, Hirota Y, Noce T (1998) In vivo gene transfer to mouse spermatogenic cells by deoxyribonucelic acid injection into seminiferous tubules and subsequent electroporation. Biol reprod 59: 1439–1444

Yoshinaga K, Nishikawa S, Ogawa M, Hayashi S, Kunisada T, Fujimoto T, Nishikawa S (1991) Role of c-kit in mouse spermatogenesis: identification of spermatogonia as a specific site of c-kit expression and function. Development 113: 689–699

7 Regulatory Mechanisms in Mammalian Spermatogenesis

D.M. de Kretser, T. Meehan, M.K. O'Bryan, N.G. Wreford,
R.I. McLachlan, K.L. Loveland

7.1 Classical Concepts of Hormonal Regulation 88
7.2 The Impact of Sertoli Cells on the Regulation of Spermatogenesis 92
7.3 Generation and Replication of the Gonocytes and Spermatogonia 94
7.4 Molecular Mechanisms Involved in the Transduction of Hormonal Signals from the Sertoli Cell to Germ Cells 96
7.5 The Role of Apoptosis in Hormonal Action 99
References 100

The output of spermatozoa by the testis is carefully regulated through a number of mechanisms that ultimately determine the proliferation and survival of germ cells. Because of the intimate relationship of the germ cells to the supporting cells, termed Sertoli cells, any review of the regulatory mechanisms cannot proceed without considering their effects on the Sertoli cells. The links between the cell types are considerable and well beyond the scope of this chapter but are covered in several reviews(Bardin et al 1994; Russell & Griswold 1993). Consequently this chapter reviews the classical concepts concerning the hormonal regulation of spermatogenesis but also considers key aspects of the physiology of the Sertoli cells. While the emphasis concerns the proliferation of germ cells, the review briefly considers the process of apoptosis because the survival of germ cells depends on the balance between these processes.

7.1 Classical Concepts of Hormonal Regulation

It is well accepted that normal spermatogenesis requires a fully functional hypothalamo-hypophysial system which generates stimulatory signals consisting of the gonadotrophic hormones, FSH and LH. Through the action of LH on the Leydig cells in the testis, high local concentrations of testosterone (T) are generated within the testis which are important for normal spermatogenesis.

Role of FSH and Testosterone in the Initiation of Spermatogenesis. These concepts are well illustrated by the parallel rise of FSH and LH during the initiation of spermatogenesis during puberty and by the importance of these two hormones in the recrudescence of spermatogenesis in seasonal breeders. Further, in men with Kallman's syndrome, the failure of GnRH secretion leads to low or undetectable levels of FSH and LH, absent pubertal maturation and the need for treatment with both gonadotrophins or GnRH to initiate spermatogenesis (Finkel et al 1985). In such patients the use of FSH and LH rarely achieves normal sperm counts and testis size, most probably due to a subnormal Sertoli cell complement associated with the failure of prenatal and neonatal stimulation by FSH (Schekter et al 1988; see below). The requirement for FSH in the initiation of spermatogenesis has been questioned as a result of several findings. In men with inactivating mutations of the FSH receptor gene, spermatogenesis proceeded to completion but sperm counts were low and testis volumes were decreased (Tapanainen et al 1997). Further the observation of complete spermatogenesis in mice with the targeted disruption of the FSH βsubunit gene (Kumar et al 1997) also called into question the requirement for FSH. However, although these mice were fertile, their testes were smaller than normal and quantitative studies indicate decreased Sertoli cell number and a complementary decrease in the number of germ cells (Wreford et al unpublished). Further, in the hpg mouse line which also has undectable levels of both FSH and LH due to the lack of GnRH, Singh et al (1995) showed that T could initiate spermatogenesis albeit with a lower sperm output. Despite this data, many patients with Kallman's syndrome cannot achieve complete spermatogenesis without the use of FSH in combination with LH reinforcing the concept that both hormones play critical roles in the establishment of normal spermatogenesis.

The Role of Testosterone in the Maintenance of Spermatogenesis. The concentrations of T in the testis are considerably higher than the circulating levels since T is produced locally through the action of LH on Leydig cells. Several studies have evaluated the concentrations of T that are essential for spermatogenesis and a consensus exists that levels representing 10–20% of normal intra-testicular concentrations will maintain relatively normal sperm output in rats (Clermont & Harvey 1965; Cunningham & Huckins 1979, Marshall et al 1983; Sun et al 1989). It is still unclear why such high concentrations of T, relative to peripheral levels, are necessary to maintain spermatogenesis. Tissues such as the seminal vesicles become enlarged when exposed to the same concentrations of T required for the maintenance of spermatogenesis despite apparently using the same AR as that found in the testis to mediate androgen action (Sun et al 1989).

The cellular site of T action is still a matter of debate. On the basis of *in vivo* studies O'Donnell et al (1996a) proposed that T was essential in preventing the premature desquamation of round spermatids from the seminiferous epithelium. In support of this concept, they found that both T and FSH stimulate the production of the adhesion molecule, N-cadherin, which may be involved in spermatid binding to Sertoli cells (Perryman et al 1996). Those studies also showed that T exerted an action on the conversion of spermatogonia to spermatocytes and the transformation of spermatocytes to spermatids (Sun et al 1990; O'Donnell et al 1996a) indicating a wider role for T in the maintenance of spermatogenesis. However, using the hpg mouse model of gonadotrophin releasing hormone deficiency, Singh et al (1995) proposed that the most T sensitive step was the conversion of spermatocytes to spermatids.

The role of dihydrotestosterone (DHT) in spermatogenesis has remained enigmatic but recent studies suggest, that in the presence of low intratesticular testosterone concentrations, DHT can assist in the maintenance of germ cells with a known T dependence (O'Donnell et al 1996b). They showed by the use of a 5α-reductase inhibitor that when intra-testicular levels were lowered by the use of 6 cm T silastic implants, the decreased DHT levels impaired the progression of round spermatids through spermiogenesis. However, when larger implants of T were used (10 and 24 cm), this effect of the decreased DHT could not be demonstrated. In a further study, O'Donnell et al (1999) showed that the use of 5α reductase inhibitors in the setting of low intratesticular T levels could successfully suppress the number of elongated spermatids arising from the progression of round spermatids through mid-spermiogenesis.

Further evidence for the important role played by T in spermatogenesis emerged from studies by Matsumoto and colleagues (1983) who demonstrated that sperm output in men could be drastically suppressed by the administration of T to normal men. This action of T was exerted by the suppression of FSH and LH, the latter leading to a decrease in intra-testicular T levels. They also showed that if hLH was given to these men while exogenous T was continued, the stimulation of Leydig cell T production partially restored sperm output (Matsumoto et al 1984). Our recent study showed that the principal lesion induced by the elevation of T to supraphysiological levels (63.2± 12 nmol/L), was a striking decrease in the number of type B spermatogonia to about 10% of normal (Zhengwei et al 1998a). Later stages of spermatogenesis were also suppressed but there was a relatively greater decline in step 3–6 spermatids in comparison to steps 7–8, suggesting a striking retention of elongated spermatids in the epithelium despite the presence of azoospermia. The latter data suggest a defect in spermiation but the underlying mechanism remains to be elucidated. This study also identified a significant heterogeneity in the degree of spermatogenic disruption in adjacent tubules, raising the question of how these differences can be maintained given that they are exposed to a common hormonal milieu. It is possible that regional differences may occur due to as yet undefined mechanisms that, for instance, maintain a higher local concentration of T in an individual tubule or to higher concentrations of growth factors that may synergize with hormonal stimuli. Alternatively, any cytological study can only assess the balance between proliferative activity and the rate of cell death due to apoptosis, a process that is only partly understood in the testis (see below). Stage dependent differences in spermatogenic development may appear exaggerated by hormonal perturbations. Finally, the biopsies obtained capture the state of the spermatogenic processes at a single time point in a treatment program extending over several months.

While the case for the importance of T in the maintenance of spermatogenesis is not disputed, further evidence can be adduced from the outcomes of mutations in the androgen receptor (AR). It is well recognized that mutations in the AR can cause total or partial androgen insensitivity leading to a range of patterns of abnormal sexual differentiation (Quigley et al 1995). However several patients have been identified in whom point mutations in the AR has resulted in spermatogenic disruption as the sole phenotype (Wang et al 1998). Further, an increase in the size of the trinucleotide repeat encoding the polyglutamine tract found in exon 1 of the AR, has been linked to spermatogenic disruption (Tut et al 1997; Dowsing et al 1999). The expansion of this CAG repeat may impair transactivation resulting from ligand binding to the receptor further emphasizing the importance of androgen action.

The Role of FSH in the Maintenance of Spermatogenesis. As indicated earlier, there is considerable debate as to whether FSH is essential for the maintenance of spermatogenesis. Some of these concerns arose from the demonstration that T could maintain spermatogenesis in rat models wherein FSH was assumed to be low (Clermont & Harvey 1965). Further studies have shown that although low doses of T suppress FSH, higher doses of T return FSH concentrations to essentially normal levels (Rea et al 1986; Sun et al 1989). In a recent study, Meachem et al (1998) showed that the neutralization of FSH *in vivo* by the use of a polyclonal antibody against FSH, could impair the action of T in restoring spermatogenesis in the testosterone-estradiol suppressed or GnRH immunised model. In the former model, FSH neutralization interfered with the ability of higher doses of T to restore spermatid numbers whereas in the latter model the restoration of spermatogonial numbers was inhibited. Other data strongly suggest that FSH acts specifically to increase spermatogonial numbers when it is used to correct gonadotrophin deficiency in rodents and primates (Meachem et al 1996; Zhengwei et al 1998b). The concept that T and FSH synergize in their actions on spermatogenesis is in accord with suggestions made by Matsumoto and colleagues (1983; 1984) from a series of studies in men.

Localization of T and FSH Receptors. Most evidence to date would support the view that the AR and the FSH receptors are not located on germ cells but can be found on the somatic cells of the testis. The AR is located in Sertoli cells, Leydig cells and peritubular cells in the testis (Bremner et al 1994). A report suggesting that AR could be found by immunocytochemistry on spermatids (Vornberger et al 1994) has not been confirmed. As indicated earlier, LH exerts its actions through specific receptors on Leydig cells by the stimulation of T (de Kretser et al 1971) but may also exert an influence by its effects on testicular blood flow through receptors on endothelial cells (Ghinea et al 1994).

FSH receptors are found on Sertoli cells and the highest levels are found on Sertoli cells in tubules at stages I-II of the cycle (Parvinen 1982). Some earlier reports suggested that labelled FSH bound to spermatogonia and Sertloi cells (Orth & Christensen 1978) but the localization to spermatogonia has not been confirmed by demonstration of FSH receptor mRNA or protein on these cells.

The absence of FSH and T receptors on germ cells and their localization on Sertoli cells emphasizes the crucial role of the somatic cells in the regulation of spermatogenesis. It is not only the site of these receptors but the close anatomical association of the Sertoli and germ cells in the epithelium that places these cells in an unique position to influence sperm production. It is not possible in this chapter to outline the many mechanisms by which the Sertoli cells could influence spermatogenesis. However, it is clear that the actions of FSH and T on Sertoli cells must be transmitted to germ cells by intermediary molecules, many of which remain to be defined and characterised.

7.2 The Impact of Sertoli Cells on the Regulation of Spermatogenesis

There are numerous aspects of Sertoli cell biology that can influence spermatogenesis and these are so extensive that a whole volume has been devoted to this cell (Russell & Griswold 1993). These range from the intimate association of Sertoli cells with germ cells and their structural role in establishing the blood-testis barrier to the importance of the cytoskeleton in providing support to the epithelium(see review by Bardin et al 1994). This section will emphasize the relationship of Sertoli cell number to the total spermatogenic output of the testis and the importance of the state of Sertoli cell "maturity" to the onset of spermatogenesis. These developmental aspects of the Sertoli cell set the foundations of spermatogenesis in fetal, neonatal and pubertal phases of testicular development.

A considerable body of data supports the concept that the number of Sertoli cells is an important determinant of total sperm output by the testis. This concept implies that each Sertoli cell can support a finite number of germ cells as they pass through the various stages of spermatogenesis. The periods of Sertoli cell proliferation occur prenatally, immediately postnatally in some species such as rodents and for differing lengths of time in species wherein birth is separated from puberty by an extended duration. Thus in man, there is data to support a prenatal phase, a post-natal phase which may extend for 6–12 months and a pubertal phase (Cortes et al 1987; Andersson et al 1998a). Since the period of proliferation occurs early in life, the supporting template within the seminiferous tubules is often well established before a boy enters the pubertal proliferative phase. In contrast, the principal period of Sertoli cell proliferation in the Rhesus monkeys was during pubertal maturation (Marshall & Plant 1996). These data indicate the important role that these developmental processes play in determining the ultimate sperm output in a mammal.

The majority of data supporting these concepts have been derived from the study of Sertoli cell division in rats, a species in which the prenatal phase extends continuously from the formation of the seminiferous cords to term and the postnatal phase terminates about day 18 after birth (Orth et al 1982). If FSH is elevated by hemicastration on day 1 to 2, increased Sertoli cell proliferation results and manifests in adulthood by compensatory hypertrophy of the remaining testis (Cunningham et al 1978). Further, rendering neonatal rats hypothyroid by the ingestion of anti-thyroid drugs, extends the period of Sertoli cell division to approximately 30 days and results in increased numbers of Sertoli cell per testis and an increased sperm output in adult life (Cooke et al 1991). The combination of neonatal hypothyroidism and unilateral castration (raising FSH), results in a further increment in Sertoli cell number and adult sperm output indicating the presence of two discrete mechanisms of hormonal control of Sertoli cell division (Simorangkir et al 1995). The levels of inhibin B in serum appears to reflect the numbers of Sertoli cells in the rat testis in a number of models in which Sertoli cell number was manipulated experimentally (Sharpe et al 1999). Similar conclusions were reached by Ramaswamy et al (1999), in a study of the effect of unilateral orchidectomy in Rhesus monkeys.

There is accumulating data to indicate that the Sertoli cell undergoes structural and functional changes during sexual maturation. For instance, the size of the Sertoli cell nucleus increases during sexual maturation probably reflecting the action of FSH (Meachem et al 1998). During a study of the effects of neonatal hypothyroidism on Sertoli cell function showed that there was a significant delay in changes of a number of biochemical parameters including the delay in the normal elevation of β_B mRNA (Bunick et al 1994). The effects of this delay on germ cell maturation was illustrated by the profound delay in the progression of gonocytes through spermatogonia to spermatocytes and spermatids (Simorangkir et al 1997). Not only was the progression delayed but the maturing germ cells appeared to show the features of apoptosis. These data highlight the importance of coordinated maturation of both Sertoli cells and germ cells, without which the completion of spermatogenesis is disrupted.

7.3 Generation and Replication of the Gonocytes and Spermatogonia

Spermatogenesis depends on the provision of a pool of renewing spermatogonia which, at regular intervals, enter the meiotic process. Spermatogonia arise from the gonocytes which in turn take their origin from the primordial germ cells that migrate into the gonadal ridge. The number of primordial germ cells, which are incorporated into the developing seminiferous cords and survive, effectively establish the potential size of the germ cell pool from which the subsequent germ cells develop. The factors which control the migration and survival of the primordial germ cells have been the subject of numerous studies and their findings are beyond the scope of this chapter (see reviews Galli et al 1994; Loveland & Schlatt 1997).

The primordial germ cells divide and differentiate to form gonocytes, sometimes termed prospermatogonia, which in the rat testis cease cell division. These cells are centrally placed within the seminiferous cords and on the 3rd day after birth, they recommence proliferation and migrate to establish contact with the basement membrane (Orth et al 1997).

The role of the conventional regulators of spermatogenesis, FSH and T, on gonocyte division, migration and transformation into spermatogo-

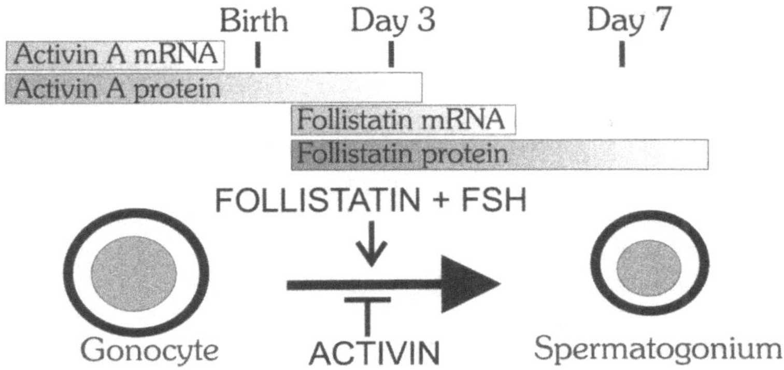

Fig. 1. This diagram illustrates the changes in activin A and follistatin in testicular germ cells during the perinatal period

nia is not well established but is likely to be achieved through their effects on Sertoli cell proliferation and maturation. Many locally produced growth factors can influence gonocyte proliferation such as platelet derived growth factor (Li et 1997), Leukaemia inhibitory factor (de Miguel et al 1996), oncostatin M (de Miguel 1997) and basic fibroblast growth factor (van Dissel-Emiliani et al 1996).

Recently we have explored the actions of activin A and its binding protein, follistatin on the gonocyte to spermatogonial transformation partly in view of the reported inhibitory effect of activin A on the proliferation of migrating primordial germ cells (Richards et al 1999). $\beta_A$subunit mRNA is expressed in gonocytes from 18.5 to 21.5 days post-coitum and the corresponding protein, probably indicating the presence of activin A, is found in these cells until day 3 postnatal when the gonocytes transform into spermatogonia. In contrast, follistatin mRNA and protein is not found in gonocytes at birth but appears from day 3 and is present in spermatogonia. The functional significance of these localizations emerged from our study of gonocytes in testicular tissue from day 3 rat testes maintained in a fragment culture system (Boitani et al 1993), where activin A inhibited the migration of gonocytes to the basement membrane of the seminiferous cords and increased gonocyte numbers (Meehan et al 2000). Follistatin neutralized these actions in keeping with its known capacity to antagonize the actions of activin A. In addition, the combined treatment with follistatin and FSH increased the numbers of spermatogonia defined as gonocytes that had established contact with the basement membrane of the cord. Further, activin A inhibited Sertoli cell proliferation as judged by the incorporation of bromodeoxyuridine. These actions, and the regulated expression of activin and follistatin in the gonocytes, implicates these proteins in the control of proliferation of these cells and their differentiation into spermatogonia. It is also possible that activin A may increase gonocyte numbers by inhibiting apoptosis. Because the Sertoli cell also has the capacity to produce activin and follistatin as well as inhibin, it is likely that these molecules are part of an integrated network of autocrine and paracrine factors that coordinate Sertoli cell and germ cell maturation (Fig. 1).

In other studies, Boitani et al (1993) showed that committed spermatogonia in the day 9 rat testis responded to activin A by proliferation, in keeping with earlier in vitro data using co-cultures of spermatogonia and Sertoli cells (Mather et al 1990).

7.4 Molecular Mechanisms Involved in the Transduction of Hormonal Signals from the Sertoli Cell to Germ Cells

As discussed earlier in this chapter, the absence of receptors for the gonadotrophic hormones and T on germ cells makes it highly probable that the Sertoli cell is the intermediary in the transmission of hormonal signals to the germ cells. One such example is the inability of the seminiferous epithelium to "retain" round spermatids in the presence of low intratesticular T concentrations (O'Donnell et al 1996a). The likely mechanism is the failure of the Sertoli cell under such hormonal conditions to maintain the levels of adhesion molecules to enable binding of round spermatids (Cameron & Muffly 1991; Perryman et al 1996). Considerable effort has been expended in attempts to isolate the molecules responsible for the "cross-talk" between Sertoli cells and germ cells but their specific nature remains unknown (Pineau et al 1993).

There is a strong possibility that a range of growth factors present in the seminiferous epithelium may be involved in the transduction of hormonal stimuli to germ cells. To be likely candidates, the proteins must be produced by the Sertoli cells and receptors for the proteins must be present on germ cells. Further, FSH or T, should modulate the production of the relevant protein. There are many growth factors that can meet the first two of these criteria but evidence for the modulation of their levels by FSH and T is lacking due to the absence of suitable assays with the sensitivity and specificity to document these changes. It is not possible to canvass the range of growth factors that can fulfill some of these criteria but some examples are provided.

The actions of stem cell factor are crucial for the progression and survival of primordial germ cells and spermatogonia, and the disruption of spermatogenesis through mutations in SCF or its receptor, c-kit, are well documented (For review see Loveland & Schlatt 1997). The injection of a monoclonal antibody to c-kit to neutralize SCF binding to c-kit led to decreased numbers of spermatogonia (Yoshinaga et al 1991) and led to greater levels of apoptosis in germ cells undergoing both mitosis and meiosis (Packer et al 1995). These data emphasised the importance of the c-kit/SCF system for the maintenance of germ cells in the adult testis. We and others showed that mRNA for SCF is up-regulated in the rat testis at the time when gonocytes differentiate to form spermatogonia (Munsie et al 1997; Yan et al 1999a), during the time when production of a membrane associated form of SCF protein is increased through an alternative mRNA splicing mechanism (Huang et al 1992). The link between the classical hormonal regulators and SCF has been strengthened by the observation that SCF levels are increased by FSH (Yan et al 1999a). Further, they have subsequently shown that this stimulation is translated into increased numbers of germ cells by the observation that the elevated SCF decreases the frequency of germ cells undergoing apoptosis (Yan et 1999b). Their study also showed that this antiapoptotic effect was exerted on spermatocytes and spermatids that express a truncated form of c-kit (Vincent et al 1998). These results provide an excellent example of how Sertoli cells transmit the actions of classical regulators of spermatogenesis to germ cells. This interaction also illustrates further the involvement of the Sertoli cells in regulating germ cell survival since, the alternative splicing event required to generate the membrane-bound form of SCF, is facilitated by low pH probably generated by the production of lactate, the preferred substrate of germ cells, by the Sertoli cells (Mauditt et al 1999).

As discussed earlier, we have demonstrated that the inhibin/activin/follistatin system is involved in the regulation of germ cell events in the early postnatal rat testis. The members of this system are also expressed in the adult testis where their local functions are not clearly defined. Messenger RNA and protein for the α, β_A, β_B subunits of inhibins/activins are located within the Sertoli cells and the testis is the principal source of inhibin in the circulation. Further, in the male the major form of inhibin in serum is inhibin B ($\alpha\beta_B$). Both β subunits are found in the Sertoli cell and Grootenhius et al (1989) have shown that activin A is present in pooled culture media from Sertoli cells. Recently, Andersson et al (1998b) have shown that the β_B subunit can be localized by immunocytochemistry to the primary spermatocytes and spermatids in the human testis and we have similar data in the rat using both immunocytochemistry and in situ hybridisation. However the effects of FSH and T stimulation on the β subunits, both in germ cells and Sertoli cells is unclear since the majority of studies have utilized Northern analysis of total RNA (Krummen et al 1989). We and others have shown that follistatin protein and mRNA is found, not only in the Sertoli cells, but also in spermatogonia, spermatocytes and round to condensing spermatids in the rat testis (Michel et al 1993; Meinhardt et al 1998). These localisations are important in considering the role played by the activins since follistatin can neutralize the majority of their actions (see review by Phillips and de Kretser 1999).

As yet the local actions of these proteins in the seminiferous epithelium are not fully understood but the activins have been shown to maintain the condensed state of mitochondria in spermatocytes (Meinhardt et al 2000). Given the involvement of activin A in the control of apoptosis in the liver and in B lymphocytes, their role in this process in the testis requires exploration (Schwall et al 1993; Brosh et al 1995)).

The regulation of whether activin or inhibin is produced at the site where both the α and β subunits are synthesized is complex. The generally held view suggests that the presence of excess α subunit will direct the cell to the production of inhibin. In this context, the known ability of FSH to stimulate α subunit mRNA in Sertoli cells will drive the cell to secrete inhibin in preference to activin (Krummen et al 1989; Hancock et al 1992). Further, dose-response studies show that FSH will stimulate α subunit secretion by Sertoli cells and may well account for the observed rise in α subunit secretion in men with severe testicular damage (de Kretser et al 1989; Anawalt et al 1996). Whether the Sertoli cells, under FSH or T secretion, produce other local regulators of germ cell activin production has not been established.

Recently, the identification of activin C has been described but the nature of its actions are unknown (Hotten et al 1995). Activin C is formed by the dimerisation of the β_C subunits and these have been described in the rat testis (Loveland et al 1996) and localized by in situ hybridisation to the Sertoli cells, spermatocytes and spermatids. This distribution is similar to the localisation of the β_B subunits in germ cells and, since the β_C subunit can form potentially inactive dimers with the β_B subunits (Mellor personal communication), this represents a further potential mechanism of regulating the actions of activin B.

7.5 The Role of Apoptosis in Hormonal Action

The occurrence of apoptosis in the testis has been recognised for sometime and represents the normal mechanism of germ cell death within the testis. There have been a number of studies showing that the withdrawal of FSH and T stimulation of the testis results in apoptosis and these have been reviewed recently (Sinha Hikim et al 1999). While these studies documented the cell types which underwent apoptosis after hormone withdrawal, they did not explore the pathways that were involved in cell death. These data have been expanded by recent studies looking at the onset of apoptosis following the use of the Leydig cell cytotoxin ethane dimethane sulphonate (EDS) which results in reduction in T levels to those found in castrate animals (Woolveridge et al 1999; Nandi et al 1999). These studies explored some of the apoptotic pathways and Woolveridge et al (1999) found an elevation of Bcl-2 and Bax as well as a decrease in Fas ligand and Fas receptor in spermatocytes and spermatids, the cells under going apoptosis. However Nandi et al (1999) concluded that the Fas system was not involved in the process.

Unfortunately neither study examined the recently described Bcl-2 family member Bcl-w (Adams & Cory 1998). Recent studies have shown that targetted disruption of this gene in mice results in progressive disruption of spermatogenesis which appears increasingly after 4–6 weeks of age and by 12–16 weeks shows the phenotype of Sertoli cell only tubules (Print et al 1998; Ross et al 1998).

It is essential that further studies are performed to determine which elements of the apoptotic machinery are regulated by hormonal processes at key phases of testicular development. Such a study is critical to our understanding of future potential targets for therapeutic interventions in infertile men and for alternative methods of contraceptive development.

References

Adams JM, Cory S (1998) the Bcl-2 protein family: arbiters of cell survival. Science 281:1322–1326

Anawalt BD, Bebb RA, Matsumoto AM, Groome NP, Illingworth JP, Mc Neilly AS, Bremner WJ (1996) Serum inhibin B leevels reflect Sertoli cell function in normal men and men with testicular dysfunction. J Clin Endocrinol Metab 81:3341–3345

Andersson A-M, Toppari J, Haavisto A-M, Petersen JH, Simell T, Simell O, Skakkebaek NE (1998a) Longitudinal reproductive hormones profiles in infants: peak of inhibin B levels in infant boys exceeds levels in adult men. J Clin Endocrinol Metab 83: 675–681

Andersson AM, Muller J, Skakkebaek NE (1998b) Different roles of prepubertal and postpubertal germ cells and Sertoli cells in the regulation of serum inhibin B levels. J Clin Endocrinol Metab 83: 4451–4458.

Bardin CW, Cheng CY, Mustow NA, Gunsalus G (1994) The Sertoli cell In, The Physiology of Reproduction 2nd ed, eds Knobil E, Neill JD, Raven Press New York, p1291–1335

Boitani C, Stefanini M, Fragale A, Morena AR (1995) Activin stimulates Sertoli cell proliferation in a defined period of rat testis development. Endocrinology 136:5438–5444

Bremner WJ, Millar MR, Sharpe RM, Saunders PTK (1994) Immunohistochemical localisation of androgen receptors in the rat testis: evidence for stage-dependent expression and regulation by androgens. Endocrinology 135:1227–1234

Brosh N, Sternberg D, Honigswachs-Sha'anani J, Lee B-C, Shav-Tal Y, Tzehoval E, Shulman LM, Toledo J, Hacham Y, Carmi P, Jiang W, Sasse J, Horn F, Burstein Y, Zipori D (1995) The plasmacytoma growth inhibitor restrictin-P is an antagonist of interleukin 6 and interleukin 11. Identification as a stroma-derived activin A. J Biol Chem 270: 29594–29600.

Bunick D, Kirby J, Hess RA, Cooke PS (1994) Developmental expression of testis messenger ribonucleic acids in the rat following propylthiouracil-induced neonatal hypothyroidism. Biol Reprod 51:706–713

Cameron DF, Muffly KE (1991) Hormonal regulation of spermatid binding. J Cell Science 100:623–633

Clermont Y, Harvey SG (1965) Duration of the cycle of the seminiferous epithelium of normal hypophysectomized and hypophysectomized-hormone treated albino rats. Endocrinology 76:80–89

Cortes D, Muller J Skakkebaek NE (1987) Proliferation of Sertoli cells during development of the human testis assessed by stereological methods. Int J Androl 10:589–596

Cooke PS, Hess RA, Porcelli J, Meisami E (1991) Increased sperm production in rats following transient neonatal hypothyroidism. Endocrinology 129:237–243

Cunningham GR, Tindall DJ, Huckins C, Means AR (1978) Mechanisms for the testicular hypertrophy which follows hemicastration. Endocrinology 102: 16–23.

Cunningham GR, Huckins C (1979) Persistence of complete spermatogenesis in the presence of low intra-testicular concentration of testosterone. Endocrinology 105:177–186

de Kretser DM, McLachlan RI, Robertson DM, Burger HG (1989) Serum inhibin levels in normal men and men with testicular disorders. J Endocr 120:517–523

de Kretser DM, Catt KJ, Paulsen CA (1971) Studies on the in vitro binding of iodinated luteinizing hormone in rats. Endocrinology 88:332–337

de Miguel MP, de Boer-Brouwer M, Paniagua R, van den Hurk R, de Rooij DG, van Dissel-Emiliani FMF (1996) Leukaemia inhibitory factor and ciliary neurotropic factor promote the survival of Sertoli cells and gonocytes in a coculture system. Endocrinology 137:1885–1993

de Miguel MP, de Boer-Brouwer M, de Rooij DG, Paniagua R, van Dissel-Emiliani FMF (1997) Ontogeny and localization of oncostatin M-like protein in the rat testis: its possible role at the start of spermatogenesis. Cell Growth and Differentiation 8:611–618

Dowsing AT, Yong EL, McLachlan RI, de Kretser DM, Trounson AO (1999) Linkage between male infertility and trinucleotide repeat expansion in the androgen receptor gene. Lancet 354: 640–643

Finkel DM, Phillips JL, Snyder PJ (1985) Stimulation of spermatogenesis by gonadotropins in men with hypogonadotropic hypogonadism. New Engl J Med 313:651–655

Galli SJ, Zsebo KM, Geissler EN (1994) The kit ligand: stem cell factor. Advances in Immunology 55:1–96

Ghinea N, Mai TU, Groyer-Picard MT, Milgrom E (1994) How protein hormones reach their target cells: Receptor-mediated transcytosis of hCG through endothelial cells. J Cell Biol 125:87–97

Grootenhuis AJ, Steenbergen J, Timmerman MA, Dorsman ANRD, Schaaper WMM, Meloen RH, de Jong FH (1989) Inhibin and activin – like activity in fluids from male and female gonads : different molecular weight forms and bioactivity/immunoactivity ratios. J Endocrinol 122: 293–301.

Hancock AD, Robertson DM, de Kretser DM (1992) Inhibin and inhibin α chain precursors are produced by immature rat Sertoli cells in culture. Biol Reprod 46, 155–161.

Hotten G, Neidhardt H, Schneider C, Pohl J (1995) Cloning of a new member of the TGFβ family: a putative new activin β_C chain. Biochem Biophys Res Commun 206:608–613

Huang E, Nocka K, Buck J, Besmer P (1992) Differential expression and processing of two cell associated forms of the c-kit ligand:Kl-1 and KL-2. Mol Biol Cell 3:349–362

Krummen LA, Toppari J, Kim WH, Morelos BS, Ahmad N, Swerdloff RS, Ling N, Shimasaki S, Esch F, Bhasin S (1989) Regulation of testicular inhibin subunit messenger ribonucleic acid levels *in vivo* : Effects of hypophysectomy and selective follicle-stimulating hormone replacement. Endocrinology 125: 1630–1637.

Kumar TR, Wang Y, Lu N, Matzuk M (1997) Follicle stimulating hormone is required for ovarian follicle maturation but not male fertility. Nature Genet 15:201–204

Li H, Papadopoulos V, Vidic B, Dym M, Culty M (1997) Regulation of rat testis gonocyte proliferation by platelet-derived growth factor and estradiol: identification of signalling mechanisms involved. Endocrinology 138:1289–1298

Loveland KL, McFarlane JR, de Kretser DM (1996) Expression of activin β C subunit mRNA in reproductive tissues. J Mol Endocr 17: 61–65

Loveland KL, Schlatt S (1997) Stem cell factor and c-kit in the mammalian testis: lessons originating from Mother Nature's gene knock-outs. J Endocr 153:337–344

Marshall GR, Wickings EJ, Ludecke DK, Nieschlag E (1983) Stimulation of spermatogenesis in stalk-sectioned rhesus monkeys by testosterone alone. J Clin Endocrinol Metab 57:152–159

Marshall GR, Plant TM (1996) Puberty occurring spontaneously or induced precociously in rhesus monkey (Macaca mulatta) is associated with a marked proliferation of Sertoli cells. Biol Reprod 54:192–1199

Mather JP, Attie K, Woodruff T, Rice G, Phillips D (1990) Activin stimulates spermatogonial proliferation in germ-Sertoli cell cocultures from immature rat testis. Endocrinology 127:3206–3214

Matsumoto AM, Paulsen CA, Bremner WJ (1984) Stimulation of sperm production by human luteinizing hormone in gonadotropin-suppressed normal men. J Clin Endocrinol Metab 59:882–887

Matsumoto AM, Karpas AE, Paulsen CA, Bremner WJ (1983) Reinitiation of sperm production in gonadotropin-suppressed normal men by the administration of follicle stimulating hormone. J Clin Invest 72:1005–1015

Mauditt C, Chatelain G, Magre S, Brun G, Benahmaed M, Michel D (1999) Regulation by pH of the alternative splicing of the stem cell factor pre-mRNA in the testis. J Biol Chem 272:770–776

Meachem SJ, McLachlan RI, de Kretser DM, Robertson DM, Wreford NG (1996) Neonatal exposure of rats to recombinant human follicle stimulating hormone increases adult Sertoli cell and spermatogenic cell numbers. Biol Reprod 54:36–44

Meachem SJ, Wreford NG, Stanton PG, Robertson DM, McLachlan RI (1998) Follicle stimulating hormone is required for the initial phase of spermatogenic restoration in adult rats following gonadotropin suppression. J Androl 19:725–735

Meehan T, Schlatt S, O'Bryan MK, de Kretser DM, Loveland KL (2000) Regulation of germ cell and Sertoli cell development by activin, follistatin and FSH. Development In Press

Meinhardt A, O'Bryan MK, McFarlane JR, Loveland KL, Mallidis C,Foulds LM, Phillips DJ, de Kretser DM (1998) Localisation of follistatin in the rat testis. J Reprod Fert 112:233–241

Meinhardt A, McFarlane JR, Seitz J, de Kretser DM (2000) Activin maintains the condensed type of mitochondria in germ cells. Mol & Cell Endo. (submitted)

Nandi S, Banerjee PP, Zirkin BR (1999) Germ cell apoptosis in the testes of Sprague Dawley rats following testosterone withdrawl by 1,2-dimethane-sulfonate administration: Relationship to Fas? Biol Reprod 61:70–75

O'Donnell L, McLachlan RI, Wreford NG, de Kretser DM, Robertson DM (1996a) Testosterone withdrawal promotes stage-specific detachment of round spermatids from the rat seminiferous epithelium Biol Reprod 55:895–901

O'Donnell L, Stanton PG, Wreford NG, Robertson DM, McLachlan RI (1996b) Inhibition of 5α-reductase activity impairs the testosterone-dependent restoration of spermiogenesis in adult rats. Endocrinology 137: 2703–2710

O'Donnell L, Pratis K, Stanton PG, Robertson DM, McLachlan RI (1999) Testosterone-dependent restoration of spermatogenesis in adult rats is impaired by a 5α-reductase inhibitor. J Androl 20:109–117

Orth JM (1982) Proliferation of Sertoli cells in fetal and postnatal rats: a quantiative autoradiographic study. Aant Rec 203:485–492

Orth JM, Christensen AK (1978) Autoradiographic localization of specifically bound ^{125}I-labelled follicle stimulating hormone on spermatogonia of the rat testis. Endocrinology 103:1944–1951

Orth JM, Gunsalus GM, Lamperti AA (1988) Evidence from Sertoli cell-depleted rats that spermatid numbers in adults depends on numbers of Sertoli cells produced during perinatal development. Endocrinology 122:787–794

Orth JM, Qui J, Jester WFJr Pilder S (1997) Expression of the c-kit gene is critical for migration of neonatal gonocytes in vitro. Biol Reprod 57:676–683

Packer AI, Besmer P, Bachvarova RF (1995) Kit ligand mediates survival of type A spermatogonia and dividing spermatocytes in postnatal mouse testes. Mol Reprod Dev 42:303–310

Parvinen M (1982) Regulation of the seminiferous epithelium. Endocrine Rev 3: 404–417.

Phillips DJ, de Kretser DM (1998) Follistatin : A multifunctional regulatory protein. Frontiers in Neuroendocrinology 19: 287–322

Print CG, Loveland KL, Gibson L, Meehan T, Stylianou A, Wreford NG, de Kretser DM, Metcalf D, Kontgen F, Adams JM, Cory S (1998) Apoptosis regulator bcl-w is essential for spermatogenesis but appears otherwise redundant. Proc Natl Acad Sci USA 95:12424–12431

Michel U, Esselman J, Nieschlag E (1993) Expression of follistatin messenger ribonucleic acid in Sertoli cell-enriched cultures: Regulation by epidermal growth factor and protein kinase A-dependent pathway. Acta Endocrinol (Copenh) 129: 525–531.

Munsie M, Schlatt S, de Kretser DM, Loveland KL (1997) Expression of stem cell factor in the postnatal rat testis. Mol Reprod Devlop 47:19–25

Perryman KJ, Stanton PJ, Loveland KL, McLachlan RI, Robertson DM (1996) Hormonal dependency of N-cadherin in the binding of round spermatids to Sertoli cells. Endocrinology 137:3877–3883

Pineau C, Syed V, Bardin CW, Jegou B, Cheng CY (1993) Germ cell-conditioned medium contains multiple factors that modulate secretion of testins, clusterin and transferin by Sertli cells. J Androl 14:87–98

Quigley CA, DeBellis A, Maischke KB, El Awady MK, Wilson EM, French FS (1995) Androgen receptor defects: historical, clinical and molecular perspectives. Endocr Reviews 16:271–321

Ramaswamy S, Marshall GR, McNeilly AS, Plant TM (1999) Evidence that in a physiological setting Sertoli cell number is the major determinant of circulating concentrations of inhibin B in the adult male rhesus monkey (Macaca mulatta). J Androl 20:430–434

Rea MA, Marshall GR, Weinbauer GF, Nieschlag E (1986) Testosterone maintains pituitary and serum FSH and spermatogenesis in gonadotrophin-releasing hormone antagonist-suppressed rats. J Endocr 108:101–107

Richards AJ, Enders GC, Resnick JL (1999) Activin and TGFβ limit primordial germ cell proliferation. Dev Biol 207:470–475

Ross AJ, Waymire KG, Moss JE, Parlow AF, Skinner MK, Russell LD, MacGregor GR (1998) Testicular degeneration in Bclw-deficient mice. Nature Genetics 18:251–256

Russell LD, Griswold MD eds (1993) The Sertoli Cell. Cache River Press, Clearwater, Florida

Schwall RH, Robbins K, Jardieu P, Chang L, Lai C, Terrell TG (1993) Activin induces cell death in hepatocytes *in vivo* and *in vitro*. Hepatology 18: 347–356.

Sharpe RM, Turner KJ, McKinnell C, Groome NP, Atanassova N, Millar MR, Buchanan DL, Cooke PS (1999) Inhibin B levels in plasma of the male rat from birth to adulthood: Effect of experimental manipulation of Sertoli cell number. J Androl 20:94–101

Sheckter CB, McLachlan RI, Tenover JS, Matsumoto AM, Burger HG, de Kretser DM & Bremner WJ (1988) Serum inhibin concentrations rise during GnRH treatment of men with idiopathic hypogonadotrophic hypogonadism. J Clin Endocrinol Metab 67:1221–1224

Simorangkir DR, de Kretser DM, Wreford NG (1995) Increased numbers of Sertoli and germ cells in adult rat testes induced by synergistic action of transient neonatal hypothyroidism and neonatal hemicastration. J Reprod Fert 104:207–213

Simorangkir DR, Wreford NG, de Kretser DM (1997) Impaired germ cell development in the testes of immature rats with neonatal hypothyroidism. J Androl 18: 186–193.

Singh J, O'Neill, Handelsman DJ (1995) Induction of spermatogenesis by androgens in gonadotropin deficient(hpg) mice. Endocrinology 136:5311–5321

Sinha Hikim AP, Swerdloff RS (1999) Hormonal and genetic control of germ cell apoptosis in the testis. Reviews of Reprod 4:38–47

Sun YT, Irby DC, Robertson DM, de Kretser DM (1989) The effects of exogenously administrated testosterone on spermatogenesis in intact and hypophysectomized rats. Endocrinology 125:1000–1010

Sun YT, Wreford NG, Robertson DM, de Kretser DM (1990) Quantitative cytological studies of spermatogenesis in intact and hypophysectomized rats: identification of androgen-dependent stages. Endocrinology 127:1215–1223

Tapanainen JS, Aittomaki K, Min J, Vaskivmo T, Huhtaniemi I (1997) Men homozygous for an inactivating mutation of the follicle stimulating hormone(FSH) receptor gene present variable suppression of spermatogenesis and fertility. Nature Genet 15:205–206

Tut TG, Ghadessy F, Trifiro MA, Pinsky L, Yong EL (1997) Long polyglutamine tracts in the AR are associated with reduced transactivation, defective sperm production and male infertility. J Clin Endocrinol Metab 82:3777–3782

Van Dissel-Emiliani FMF, de Boer-Brouwer M, de Rooij DG (1996) Effect of fibroblast growth factor-2 on Sertoli cells and gonocytes in coculture during the perinatal period. Endocrinology 137:647–654

Vincent S, Segretain D, Nishikawa S, Nishikawa S-I, Sage J, Cuzin F, Rassoulzadegan M (1998) Stage-specific expression of the kit receptor and its ligand (KL) during male gametogenesis in the mouse: a kit-KL interaction critical for meiosis. Development 125:4585–4593

Vornberger W, Prins G, Musto NA, Suarez-Quian CA (1994) Androgen receptor distribution in rat testis: new implications for androgen regulation of spermatogenesis. Endocrinology 134:2307–2316

Wang QI, Ghadessy FJ, Trounson A, de Kretser DM, McLachlan R, Ng SC, Yong EL (1998) Azoospermia associated with a mutation in the ligand-binding domain of the androgen receptor displaying normal ligand binding, but defective transactivation. J Clin Endocrin Metab 83: 4303–4309.

Woolveridge I, de Boer-Brouwer M, Taylor MF, Teerds,KJ, Wu FCW, Morris ID (1999) Apoptosis in the rat spermatogenic epithelium following androgen withdrawal: Changes in apoptosis-related genes. Biol Reprod 60:461–470

Yan W, Linderborg J, Souminen J, Toppari J (1999a) Stage-specific regulation of stem cell factor gene expression in the rat seminiferous epithelium. Endocrinology 140: 1499–1504.

Yan W, Suominen J, Toppari J (1999b) Stem cell factor protects germ cells from apoptosis in vitro. J Cell Sci 113:161–168.

Yoshinaga K, Nishikawa S, Ogawa M, Hayashi S, Kunisada T, Fujimoto T, Nishikawa S (1991) Role of c-kit in mouse spermatogenesis: identification of spermatogonia as a specific site of c-kit expression and function. Development 113:689–699

Zhengwei Y, Wreford NG, Royce P, de Kretser DM, Mclachlan RI (1998a) Stereological evaluation of human spermatogenesis after suppression by testosterone treatment: Heterogeneous pattern of spermatogenesis impairment. J Clin Endocrinol Metab 83:1284–1291

Zhengwei Y, Wreford NG, Schlatt S, Weinbauer GF, Nieschlag E, McLachlan RI (1998b) GnRH antaginist-induced gonadotropin withdrawal acutely and specifically impairs spermatogonial development in the adult macaque (Macaca fascicularis). J Reprod Fert

8 Routes of Transcriptional Activation in the Testis: CREM and its Co-Activator ACT

D. De Cesare, G. M. Fimia, A. Morlon, P. Sassone-Corsi

8.1 Introduction 107
8.2 A Multigene Family of Transcription Factors 108
8.3 Mechanisms of Activation: the Classical Scenario 109
8.4 Mechanisms of Activation: ACT, a Testis-Specific Co-Activator 115
8.5 FHL4, Another Testis-Specific LMO Protein 122
8.6 Conclusion 123
References 124

8.1 Introduction

Coordinated gene expression programs govern the complex processes of cell growth and differentiation. The modulation of gene expression by specific signal transduction pathways enables cells to trigger the appropriate short- and long-term adaptation programs in response to environmental cues. Many transcription factors are final targets of specific transduction pathways. Factors of the CREB (cAMP-responsive element binding protein) family were originally identified as activators that respond directly to the cyclic AMP-dependent signaling pathway via phosphorylation by the protein kinase A (PKA) (Montminy 1997; Sassone-Corsi 1995). This family comprises a large number of proteins encoded by the CREB, CREM (cAMP-responsive element modulator) and ATF-1 (activating transcription factor 1) genes. Recently, it has become apparent that members of the CREB family play important roles in the nuclear responses to a variety of external signals. CREB and CREM have also been shown to function in many physiological systems, including memory and long-term potentiation (Silva et al. 1998), circadian rhythms (Foulkes et al. 1997), pituitary function (Struthers et al. 1991) and spermatogenesis (Sassone-Corsi 1998).

The CREB and CREM transcription factors are activated by phosphorylation of a key serine residue by kinases stimulated by cyclic AMP, calcium, growth factors and stress signals (De Cesare et al. 1999). Phosphorylation allows recruitment of CBP (CREB Binding Protein), a large co-activator with histone acetyl-transferase activity that contacts the general transcriptional machinery. Studies of the physiological roles played by CREB and CREM have uncovered novel routes of transcriptional activation. For example, in male germ cells, CREM is not phosphorylated but associates with ACT (Activator of CREM in Testis), a member of the LIM-only class of proteins that has intrinsic transcriptional activity. Thus, in some circumstances, CREM can bypass the classical requirements for phosphorylation and association with CBP (De Cesare et al. 1999).

8.2 A Multigene Family of Transcription Factors

The ATF/CREB family includes several members in mammals: the products of the CREB, CREM and ATF-1 genes. These have been shown to be direct targets of intracellular signaling as they are phosphorylated by the cAMP-dependent protein kinase A (PKA) (Sassone-Corsi 1995). Furthermore, each one of these genes is thought to encode many isoforms, which provide additional complexity. Unique genes encode cAMP-responsive factors in *Aplysia*, *Hydra* and *Drosophila melanogaster*; these probably represent evolutionarily precursors of a gene that was then duplicated in higher eukaryotes (De Cesare et al. 1999).

CREB, CREM and ATF-1 belong to the basic domain-leucine zipper (bZip) transcription factor class of proteins and act as dimers (Sassone-Corsi 1995). They are able to heterodimerize with each other but only in certain combinations. Indeed, a "dimerization code" exists which seems to be a property of the leucine zipper structure of each factor. Dimers bind to regulatory sequences identified in the promoter elements of several genes (Lalli and Sassone-Corsi 1994). The best characterised of these sequences is the CRE (cAMP responsive element), whose consensus site is constituted by the canonical 8 bp palindromic sequence TGACGTCA (Sassone-Corsi et al. 1988). Several genes which are regulated by a variety of endocrinological stimuli contain similar sequences in their promoter regions although at different positions. CRE-binding proteins may act as both activators and repressors of transcription. The activators mediate transcriptional induction upon their phosphorylation by PKA or other kinases (see also below; de Groot et al. 1993; Gonzalez and Montminy 1989; Rehfuss et al. 1991; Sassone-Corsi 1995) and through an activation domain. Their expression is constitutive and widely distributed in various tissues in a housekeeping fashion. Among the repressors, the cAMP-inducible ICER (Inducible cAMP Early Repressor) product deserves special mention. It is generated from a cAMP-inducible alternative promoter of the CREM gene in a rapid and transient fashion in response to cAMP (Molina et al. 1993). Thus, ICER is an early response CRE-binding factor and is involved in the dynamics of cAMP-responsive transcription (Lamas and Sassone-Corsi 1997).

8.3 Mechanisms of Activation: the Classical Scenario

8.3.1 Role of Phosphorylation

Transcriptional activation by CREB and CREM requires a specific phosphorylation event that turns these proteins into powerful activators. In CREB and CREM, the phosphorylation site (Ser133 in CREB and Ser117 in CREM; see Fig. 1) is within a highly conserved region of the so-called P-box (Phosphorylation box; Radhakrishnan et al. 1997). In the classical situation Ser133 and Ser117 are phosphorylated by PKA (de Groot et al. 1994; Gonzalez and Montminy 1989) which is itself regulated by changes in intracellular cAMP levels controlled primarily by adenylyl cyclase. This enzyme is in turn modulated by various extracellular stimuli mediated by receptors and their interaction with G proteins (McKnight et al. 1988). The binding of a specific ligand to a receptor results in the activation or inhibition of the cAMP-dependent pathway, ultimately affecting the transcriptional regulation of various

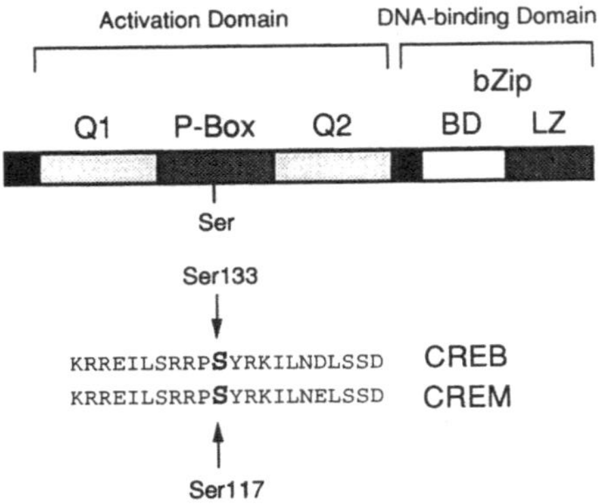

Fig. 1. Structure of activators CREB and CREM. The two proteins have a highly similar organization. The glutamine-rich domains (Q1 and Q2) and the bZip region (BD and LZ) are indicated in addition to the P-box. The amino acid sequence of the area of the CREM and CREB P-box containing Ser-117 and Ser-133 is shown. Phosphorylation at this serine residue turns CREB and CREM into activators through the interaction with the co-activator CBP. Ser-133 and Ser-117 have been shownto be phoshoacceptor sites for various kinases

genes through distinct promoter responsive sites. Increased cAMP levels directly affect the function of the tetrameric PKA complex. Binding of cAMP to two PKA regulatory subunits releases the catalytic subunits. These are translocated from cytoplasmic and Golgi complex anchoring sites and phosphorylate a number of cytoplasmic and nuclear proteins on serines in the context X-Arg-Arg-X-Ser-X (McKnight et al. 1988; Roesler et al. 1988).

Several lines of evidence now support the notion that CREB and CREM can be phosphorylated by several kinases other than PKA, these being activated by a variety of signals (De Cesare et al. 1998; de Groot et al. 1994; Ginty et al. 1994; Sheng et al. 1991; Xing et al. 1996). Importantly, the same PKA phosphoacceptor sites (Ser133 in CREB, and Ser117 in CREM) are targeted by the other kinases. For example, phosphorylation of CREB can be also elicited upon triggering of mitogenic signaling pathways by growth factors, including nerve growth factor (NGF) and epidermal growth factor (EGF) (De Cesare et al. 1998; Xing et al. 1996). The tyrosine kinase receptors for these growth factors, once activated, induce a phosphorylation cascade involving Ras, the MAP kinase kinases (MEKs), the MAP kinases ERK-1 and ERK-2 and the ribosomal S6 kinase pp90^{rsk} (Rsk) (Cohen 1997). The use of cells expressing a dominant negative Ras mutant has established that this pathway is involved in NGF-mediated phosphorylation of CREB (Ginty et al. 1994). Although CREB is not a direct substrate of MEKs or ERKs, the Rsk-2 kinase, one of three isoforms of pp90^{rsk}, phosphorylates CREB in response to EGF and NGF (De Cesare et al. 1998; Xing et al. 1996). In particular, by the use of Rsk-2-deficient human fibroblasts derived from patients affected by the Coffin-Lowry syndrome, it has been demonstrated that loss of Rsk-2 activity impairs CREB phosphorylation and consequent transcriptional induction of the c-*fos* immediate early response gene by EGF (De Cesare et al. 1998). Interestingly, p70^{s6k}, another member of the RSK family, which is activated by serum through a signaling cascade different from the ERK/MAP kinase pathway, also phosphorylates CREB (de Groot et al. 1994).

Alternative MAP kinase pathways can also lead to CREB phosphorylation. These pathways involve the p38 and the MAPKAP-2 kinases, which are targets of fibroblast growth factor (FGF) and tumor necrosis factor (TNF)-activated signaling cascades (Tan et al. 1996). Importantly, the same pathway is activated by a class of stress signals, including UV light and heat shock, to induce CREB phosphorylation at Ser133 (Iordanov et al. 1997; Tan et al. 1996). Recently, a novel family of CREB kinases has been identified, the mitogen and stress activaded kinases (MSKs), that are activated by both mitogenic and stress stimuli (Deak et al. 1998). Taken together, these results demonstrate that CREB and CREM are pleiotropic factors that act as targets of several different signaling pathways.

8.3.2 The Activation Domain

The structure of the transcriptional activation domain in CREB and CREM is basically identical. Ser-133 and Ser-117 are located in a domain identified as P-box which contains other phosphoacceptor sites for various kinases (de Groot et al. 1993; Sassone-Corsi 1995). The role of phosphorylation at these additional sites is not fully understood and appears to be secondary to Ser-133 and Ser-117. Two domains identified as Q1 and Q2 flank the P-box; they contain about three-times more glutamine residues than the remainder of the protein (Fig. 1). Glutamine-rich domains have been characterized in other factors, such as AP-2 and Sp1 (Courey and Tjian 1988; Williams et al. 1988) , where they function as transcriptional activation domains. The current notion is that they constitute surfaces of the protein which can interact with other components of the transcriptional machinery, such as RNA polymerase II cofactors. The Q2 domain appears to make a more significant contribution to the transactivation function than Q1. This is demonstrated by the properties of the two naturally occuring CREM isoforms CREMτ1 and CREMτ2 (Laoide et al. 1993) and artificially generated deletion mutants of CREB (Brindle et al. 1993). CREMτ1 and τ2 incorporate singly the Q1 and Q2 domains, respectively, CREMτ2 being a stronger transcriptional activator upon phosphorylation at Ser-117 (Laoide et al. 1993). In agreement with these results on CREM, deletion of the Q2 region in CREB dramatically abolishes activation function (Brindle et al. 1993). Furthermore, ATF-1 lacks a counterpart of the Q1 domain, and still functions as an efficient transcription activator (Hai et al. 1989). Thus, the P-box and the Q2 are sufficient to mediate cAMP-induced transcription. However, it is apparent that the activation domain is inherently a modular structure. Indeed, each component is encoded by an individual exon, so that differential splicing – as it occurs in CREM – results in the generation of factors with different activating properties (Laoide et al. 1993).

8.3.3 CBP: the Link to the Transcriptional Machinery

What are the molecular mechanisms by which phosphorylation at Ser-133 turns CREB into a transactivator? Additional proteins, known as co-activators, interact with CREB only when this is phosphorylated, directly linking signaling to the transcriptional apparatus (Arany et al. 1995; Arias et al. 1994). Indeed these co-activators, CBP and p300, are thought to interact with general transcription factors (Shikama et al. 1997). CBP and p300 are large, closely related, proteins containing three cysteine/histidine-rich domains, a bromo-domain and a glutamine-rich domain at the C-terminus (Shikama et al. 1997). CBP was isolated as a protein interacting with CREB, whereas p300 originally was identi-

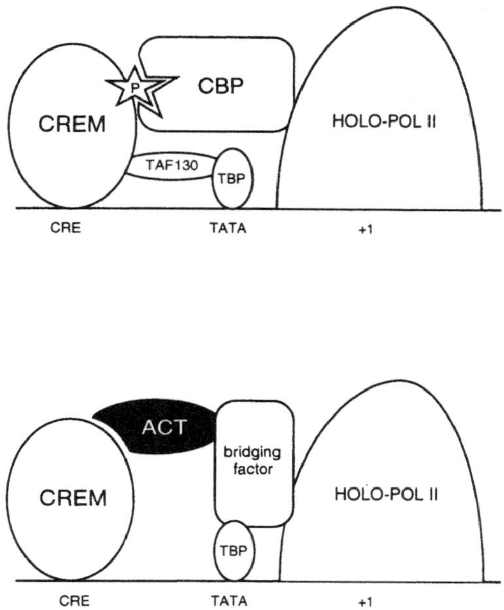

Fig. 2. CREM-mediated transcription is promoted by interaction with different co-activators. Top: a schematic representation of interaction between CREM and CBP. Phosphorylation at Ser117 (P) is required for binding to CBP and subsequent transcriptional activation, while interaction with TAF130 is constitutive. Bottom: interaction with the co-activator ACT can occur in absence of Ser117 phosphorylation. An hypothetical bridging factor, linking ACT to the basal transcription machinery is represented

fied as one of the cellular proteins associated with the adenovirus E1 A oncoprotein (Chrivia et al. 1993; Eckner et al. 1994). Phosphorylation of Ser133 promotes binding of the P-box to CBP (Fig. 2) through a region referred to as KIX (Radhakrishnan et al. 1997); this domain is highly conserved in CBP homologs from *Caenorhabditis elegans* to *Drosophila melanogaster* (Shikama et al. 1997). NMR studies have shown that the phosphorylated Ser133 participates directly in binding to the KIX (Radhakrishnan et al. 1997).

Although binding to CBP is essential, it is not sufficient for transactivation. Experiments on CREM and CREB mutants indicate that at least one of the Q-rich domains is required (Laoide et al. 1993; Montminy 1997; Sassone-Corsi 1995). The two Q-rich domains are not functionally equivalent: Q2 appears to contribute to transactivation more than does Q1 (Laoide et al. 1993). The importance of the Q-rich regions is manifest by their ability to interact directly with proteins of the basal transcriptional apparatus. Indeed, Q2 binds to the TBP-associated factor hTAF130, a subunit of the TFIID complex (Ferreri et al. 1994). Whereas binding of the P-box to CBP is inducible, because it is phosphorylation-dependent, interaction between Q2 and hTAF130 is constitutive (Ferreri et al. 1994). However, association of CREB and CREM with both CBP and TAF130 is required for efficient transcriptional activation: CREM isoforms containing only the P box or the Q2 domain behave as transcriptional repressors (Laoide et al. 1993; Shikama et al. 1997).

Both CBP and p300 display pleiotropic functions. They participate in a variety of cellular processes, such as cell growth, differentiation, DNA repair and apoptosis, by interacting with a diverse collection of transcription factors (Shikama et al. 1997). The ability of p300 and CBP to interact with various signal-responsive transcription factors indicates that they function as molecular integrators that coordinate complex signaling events at the transcriptional level. Several mechanisms by which CBP and p300 contribute to transcriptional control must therefore exist. One of them is direct recruitment of the RNA polymerase II by interaction with components of the basal transcriptional machinery, such as TFIIB, TBP (Eckner et al. 1994; Kwok et al. 1994) and RNA helicase A (Nakajima et al. 1997). Thus, CBP and p300 might function as physical bridges linking DNA-bound activators to the basal transcriptional apparatus (Fig. 2).

Another intriguing feature of CBP and p300 is their intrinsic histone acetyltransferase activity (HAT), which is thought to facilitate transcription by directly participating in chromatin remodeling at the level of inducible promoters. Importantly, CBP and p300 also regulate the acetylation of activators and basal transcription factors, thereby modulating their functions (Kouzarides 1999).

8.4 Mechanisms of Activation: ACT, a Testis-Specific Co-Activator

8.4.1 CREM Expression During Spermatogenesis

An important feature of CREB and CREM is their ubiquitous and low level of expression in all tissues (Sassone-Corsi 1995). There is however a notable exception: CREM expression in the male germ cells. CREM is the subject of a developmental switch in expression as it is highly abundant in adult testis while in prepubertal animals it is expressed at very low levels (Foulkes et al. 1992; Sassone-Corsi 1997). By a process of alternative splicing of the exons encoding the activation domain, different CREM isoforms are expressed at different times during the differentiation program of the germ cells. The abundant CREM transcript in the adult encodes exclusively the activator form, while in prepubertal testis only the repressor forms are detected at low levels. Thus, the CREM developmental switch also constitutes a reversal of function (Foulkes et al. 1992).

A remarkable aspect of the CREM developmental switch in germ cells is constituted by its exquisite hormonal regulation. The spermatogenic differentiation program is under the tight control of the hypothalamic-pituitary axis (Jégou 1993). The regulation of CREM function in testis seems to be intricately linked to FSH (follicle-stimulating hormone) signaling both at the level of the control of transcript processing and at the level of protein activity (Foulkes et al. 1993). The hormonal induction of CREM transcript levels by FSH is not transcriptional. By a mechanism of alternative polyadenylation, AUUUA destabilizer elements present in the 3' untranslated region of the gene are excluded, dramatically increasing the stability of the CREM transcript.

8.4.2 CREM is Essential for Haploid Gene Expression

A first hint as to the role of CREM during spermatogenesis was provided by its protein expression pattern. In the seminiferous epithelium CREM transcripts accumulate in spermatocytes and spermatids, but CREM protein is mainly detected in haploid spermatids (Delmas et al. 1993). This high expression of the activator protein in spermatids coincides with the transcriptional activation of several genes containing a CRE motif in their promoter region. These genes encode mainly structural proteins required for spermatozoa differentiation, suggesting a role for CREM in the activation of genes required for the late phase of spermatid differentiation. Various genes, such as RT7 (Delmas et al. 1993), transition protein-1 (Kistler et al. 1994), angiotensin converting enzyme (Zhou et al. 1996), CYP51 (Rozman et al. 1999) and calspermin (Sun et al. 1995) have been shown to be CREM targets by various experimental approaches, including *in vitro* transcription experiments with germ cells nuclear extracts. These experiments indicated that CREM participates in testis- and developmental-specific regulation of post-meiotic genes during spermiogenesis.

Genetic evidence has demonstrated that CREM is absolutely required for post-meiotic gene expression. We have generated mutant mice with targeted disruption of the CREM gene by homologous recombination (Nantel et al. 1996). Comparison of the homozygous CREM-deficient mice with their normal littermates revealed a reduction of 20–25% in testis weight and a complete absence of spermatozoa. The homozygous males are sterile. Spermatogenesis is interrupted at the stage of very early spermatids. Neither elongating spermatids, nor spermatozoa, are observed, while somatic Sertoli cells appear to be normal. This demonstrates the necessity of a functional CREM transcription factor for male fertility.

The analysis of the expression of various putative CREM target genes confirms the key role played by this transcription factor in the activation of genes such as protamine 1 and 2, transition proteins 1 and 2 and calspermin. The lack of expression of these genes may explain the impairement in the structuring of mature spermatozoa in the CREM-deficient mice (Nantel et al. 1996).

8.4.3 ACT, a Tissue-Specific Partner of CREM

The high abundance of the CREM activator in testis and its role in regulating the expression of post-meiotic genes beg the question on the mechanisms by which it exerts its function. Analysis of the phosphorylation state of CREM at various stages of the spermatogenic differentiation cycle revealed a surprising pattern: at the time CREM transcription-

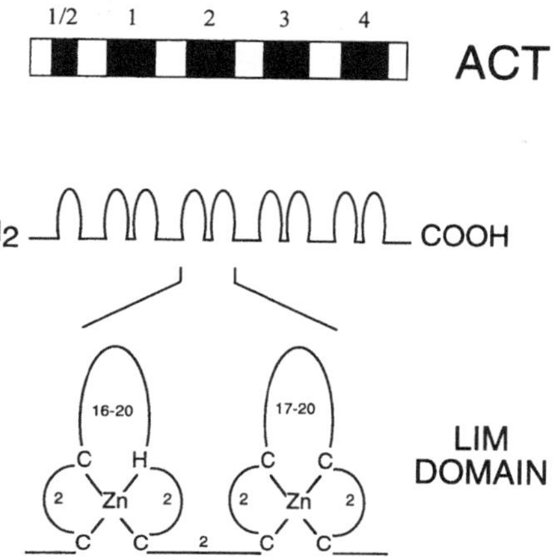

Fig. 3. Schematic representation of the structure of ACT. The repeat of LIM domains is shown on the top, each LIM domain being a black box. Each LIM domain is constituted by a double zinc finger motif, so that ACT is basically constituted by a repetition of nine zinc fingers. The double zinc-finger motif corresponding to the LIM domain is represented in the lower part

ally activates post-meiotic genes, it is unphosphorylated (L. Monaco and N. S. Foulkes, personal communication). This notion strongly suggested that activation by CREM in male germ cells may occur independently of Ser117 phosphorylation and, therefore, of the binding of CBP. We performed a yeast two-hybrid screen of a testis-derived cDNA library, using the CREM activation domain as bait, to search for a putative partner of CREM which could modulate its function in testis. We identified a clone encoding a novel protein of 284aa, named ACT (Activator of CREM in Testis) (Fimia et al. 1999). The distinctive feature of ACT is the presence of four complete LIM domains and one amino-terminal half LIM motif (Fig. 3). The LIM domain is a conserved cysteine- and histidine-rich structure of two adjacent zinc fingers, first identified in homeodomain transcription factors and subsequently found in a variety of proteins with different functions (Dawid et al. 1998). This structural motif has been shown to function as a protein-protein interaction domain (Schmeichel and Beckerle 1994). LIM domains can be present with other functional protein motifs, such as homeobox and kinase domains, but ACT belongs to the class of the LIM-only proteins (LMO), because it contains no other structural motif.

Databank searches for sequence comparison revealed that ACT has a high degree of homology with a family of proteins, whose members (DRAL/FHL-2/SLIM3, SLIM2, SLIM1/KyoT1) are expressed in heart and skeletal muscle (Chan et al. 1998; Genini et al. 1997; Morgan and Madgwick 1996; Taniguchi et al. 1998). Namely, ACT shares, with respect to the aminoacid sequence, 60% identity and 80% similarity with DRAL, a protein of unknown function expressed in heart (Genini et al. 1997).

8.4.4 Coordinate Expression of CREM and ACT

Using a specific antisense riboprobe in RNAse protection analysis, we determined the expression pattern of ACT in a variety of mouse tissues. ACT is abundantly and exclusively expressed in testis (Fig. 4; Fimia et al. 1999). Remarkably, we did not detect any signal in heart and muscle, where the DRAL, SLIM2, SLIM1 genes are expressed. We also performed a collection of analyses to identify the population of testicular cells where ACT is expressed. *In situ* hybridization studies indicate that ACT is expressed specifically in the inner rim of the seminiferous tubule (Fimia et al. 1999); expression varies depending on the stage of the tubule, which indicates a regulation during differentiation. This result was confirmed by the use of an anti-ACT specific antibody which revealed the presence of a protein of the expected size (33 kD) that accumulates at high levels in spermatid cells (Fimia et al. 1999). Immunohistochemical analysis of testis sections confirmed that ACT is present in round and elongated spermatids and showed its nuclear localisation.

CREM and ACT are co-localized in spermatids and follow the same expression pattern during testis development. During the first wave of spermatogenesis, germ cells are synchronized in their development and the temporal appearance of different cell types is well characterised. The levels of ACT transcript dramatically increased between the third and the fourth week after birth in the mouse (Fimia et al. 1998). This period corresponds, during testis development, to the end of the meiosis and the accumulation of spermatid cells. Comparative shows that ACT and CREM have overlapping expression patterns after birth (see also Fig. 5).

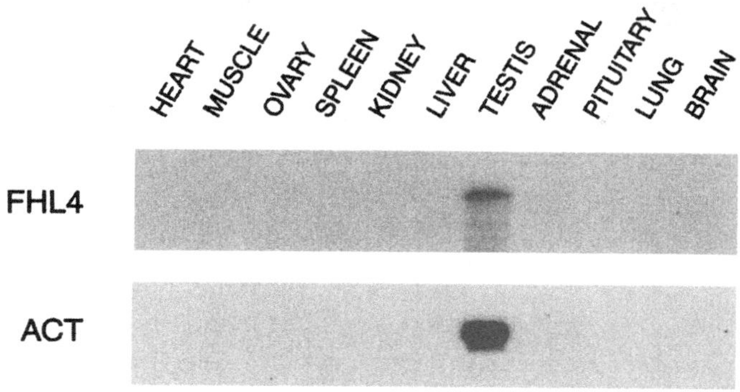

Fig. 4. ACT is exclusively expressed in testis. 10 µg of total RNA from the indicated mouse tissues were used in each lane and expression was measured by quantitative RNase protection assays. The expression of another LMO protein, FHL4, is also shown. FHL4 is also exclusively expressed in testis, but it does not interact with CREM and it does not function as a transcriptional activator

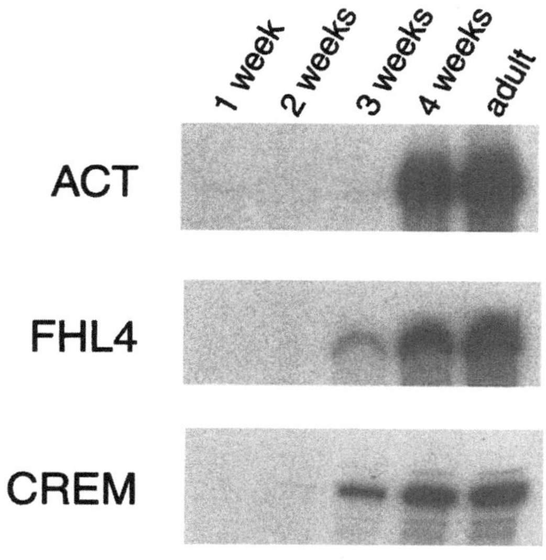

Fig. 5. Expression of ACT, FHL4 and CREM during testis development. RNAs were extracted from testes of mice at different ages (as indicated) and analysed by RNase protection assay, using specific riboprobes. 10 µg of total RNA from mouse testis were used in each lane

8.4.5 Association of CREM and ACT Proteins

The two-hybrid approach and the co-localized expression of CREM and ACT strongly suggested that the two proteins physically interact. Additional experiments demonstrate that CREM and ACT associate both *in vitro* and *in vivo*. After generation of a purified ACT-glutathione-S-transferase (GST) fusion protein, and by using various CREM deletion proteins, we have found that the P-box domain of CREM is necessary and sufficient for the association (Fimia et al. 1999). Similarly, we have generated a collection of ACT mutants where the LIM domains have been progressively deleted from either the N- and C-termini. Results obtained in GST pull-down assays revealed that the third LIM domain of ACT is necessary for efficient interaction with CREM (Fimia et al. 1999). Importantly, the CREM-ACT association occurs also *in vivo*, as demonstrated upon co-expression of the two proteins in mammalian cells followed by co-immunoprecipitation (Fimia et al. 1999). Thus, at least three lines of evidence demonstrate that the CREM and ACT proteins interact: the two-hybrid in yeast, the *in vitro* GST pull-down experiments and the *in vivo* co-immunoprecipitation in mammalian cells.

8.4.6 ACT is a Powerful Co-Activator

Sequence analysis of ACT cDNA obtained from the two-hybrid screening revealed, surprisingly, that the ORF of ACT was not in frame with the sequence encoding the GAL4 activation domain within the yeast expression vector. This observation suggested that the ACT protein could be translated from its own AUG and, therefore, have intrinsic properties of a coactivator. To investigate this possibility we tested a series of constructs where ACT was placed in combination or not with the Gal4 activation domain. We established that ACT, *per se*, has a potent activation capacity as it turns the inactive CREM into a powerful transcriptional activator (Fimia et al. 1999). Analogous results have been obtained with CREB. Our results also demonstrate the presence of an autonomous activation domain within the protein. Indeed, by fusing ACT with the Gal4 DNA-binding domain we have shown that its recruitment to DNA is sufficient, *per se*, to elicit transcriptional activation.

Is the potent activation function of ACT capable to turn a repressor into an activator? We have previously described that the CREM gene encodes various isoforms, which behave both as activators (CREMτ, CREMτ1, CREMτ2) and repressors (CREMα, CREMβ, CREMγ) (Laoide et al. 1993; Sassone-Corsi 1995). Therefore, we investigated whether ACT could elicit transactivation even when coupled to one of the repressor isoforms. We have found that a CREM isoform lacking the glutamine-rich domains, but keeping the P-box (i. e. CREMα; (Foulkes et al. 1991), is not able to drive transcription in presence of ACT, although it can associate with ACT *in vitro*. Additional results show that the Q2 domain, but not the Q1, has some transcritpional activity when co-expressed with ACT (Fimia et al. 1999). Thus, ACT-mediated induction of transcription takes place only in the presence of CREM activator isoforms. The presence of the P-box in combination with at least one glutamine-rich domain seems to be required.

These results supported the view that in male germ cells ACT would provide the activation function which is lacking by the absence of CREM phosphorylation in Ser117. Indeed, in yeast where both CREB and CREM are inactive because of the lack of CBP, the co-expression of ACT with these proteins results in a powerful transcriptional activation. Additional experiments in F9 and COS cells indicate that ACT is also able to co-activate both CREB and CREM upon transfection (Fimia et al. 1999). Thus, ACT functions as a powerful co-activator in mammalian cells.

8.4.7 Phosphorylation- and CBP-Independent Activation

The absence of CREM phosphorylation at the time of transcriptional activation of post-meiotic genes suggested that ACT may be able to exert its function on a CREM protein with the Ser117 mutated. As mentioned above, phosphorylation at Ser117 in CREM and at Ser133 in CREB is necessary for association with CBP and transcriptional activation (Chrivia et al. 1993). We tested, in the two-hybrid assay, a CREM mutant protein bearing a serine to alanine substitution at position 117, a mutation which prevents phosphorylation of the protein at this site (de Groot et al. 1994). A similar degree of activation by ACT is observed with the wild type CREM protein and the Ser117Ala mutant, demonstrating that CREM phosphorylation is not required for ACT to exert its function (Fimia et al. 1999). Thus, as demonstrated by the experiments in yeast, ACT is a co-activator whose action is CBP-, TAF130- and Ser133/117 phosphorylation-independent (see Fig. 2).

Our results reveal that transcriptional activation by CREB and CREM may be obtained through at least two different molecular pathways. The first, classical scenario, involves phosphorylation at Ser133/Ser117, subsequent association with CBP and interaction of TAF130 with the Q2 region. The second, which occurs in a cell-specific manner, involves ACT and seems to bypass the above mentioned requirements (Fig. 2).

8.5 FHL4, Another Testis-Specific LMO Protein

A large number of LIM proteins have been implicated in various steps of development and cellular differentiation programs. As mentioned above, LIM domains can be present with other functional protein motifs, such as homeobox and kinase domains, but ACT belongs to the class of LMO (LIM-only) proteins because it contains no other structural motif. Several other proteins belong to this class, containing a variable number of LIM domains. However, as we have shown (Fimia et al. 1998), there is high functional specificity as for instance LMO-2 - a protein containing only two LIM domains – is not able to activate transcription as ACT does.

A small group of LMO proteins has been identified as the FHL sub-class. These are members of the LIM family defined by their particular secondary arrangement of LIM domains. As ACT, they consist of four LIM domains and an N-terminal single zinc finger motif (hence Four-and-a-Half-LIM proteins).

Recently, the cDNA sequence encoding for a novel member of the FHL family, FHL4, has been reported. FHL4 expression seems to be testis-specific (Morgan and Madgwick 1999). We have performed RNase protection analyses with a specific riboprobe for FHL4 to establish its expression pattern using a panel of adult tissues. Interestingly, as ACT, FHL4 is exclusively present in the testis (see Fig. 4; unpublished results). Moreover, analysis of FHL4 expression during postnatal development of the testis shows that transcripts are detected from the third

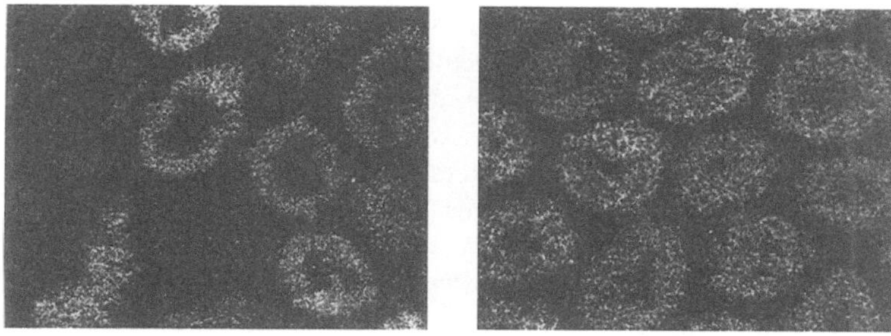

Fig. 6. Histological analysis of ACT and FHL4 transcripts distribution in adult mouse testis by *in situ* hybridization using specific probes. FHL4 is more widely expressed than ACT. CREM expression overlaps perfectly with ACT (Fimia et al. 1998)

week, suggesting that FHL4 is expressed in male germ cells (see Fig. 5; unpublished results). Indeed, in situ hybridization experiments confirm that the FHL4 expression is restricted to specific seminiferous tubules, although with a somewhat wider pattern than ACT (Fig. 6). This pattern of expression, apparently overlapping with ACT, prompted additional studies on the putative regulatory function of FHL4. Strikingly, especially considering the relatively high degree of similarity between ACT and FHL4, we have found no functional interaction between CREM and FHL4 in the two-hybrid assay, nor physical association as tested by *in vivo* co-immunoprecipitation (unpublished results). In addition, FHL4 has no intrinsic transcriptional activation potential. Taken together, these results indicate that FHL4 and ACT have different roles in the development of the testis and in the regulation of post-meiotic genes. FHL4 constitutes an interesting tool to the identification of its natural partner in male germ cells.

8.6 Conclusion

The identification of ACT opens new horizons, showing that activation of CREM- and CREB-mediated transcription can be CBP independent and phosphorylation independent. How general is this mechanism? The expression of ACT is restricted to adult male germ cells, but as mentioned here other ACT-like molecules have been described. Among these, DRAL (Genini et al. 1997), another FHL protein, shares 80% similarity with ACT but is expressed in different tissues, mainly the heart. Importantly, also DRAL interacts specifically with members of the CREB family. Thus, the presence of proteins that are related to ACT and have diverse tissue-specific patterns of expression suggests that a novel class of activators operating within specific differentiation programs exists. The restricted expression of ACT raises also the question about its specific role during the spermatogenesis and its influence on the differentiation process of the germ cells. The generation of mutant mice with targeted deletion of the ACT gene will help to elucidate this question.

References

Arany Z, Newsome D, Oldread E, Livingston DM, Eckner R (1995) A family of transcriptional adaptor proteins targeted by the E1 A oncoprotein. Nature 374: 81–84

Arias J, Alberts AS, Brindle P, Claret FX, Smeal T, Karin M, Feramisco J, Montminy M (1994) Activation of cAMP and mitogen responsive genes relies on a common nuclear factor. Nature 370: 226–229

Brindle P, Linke S, Montminy MR (1993) Protein-kinase A-dependent activator in transcription factor CREB reveals new role for CREM repressors. Nature 364: 821–824

Chan KK, Tsui SK, Lee SM, Luk SC, Liew CC, Fung KP, Waye MM, Lee CY (1998) Molecular cloning and characterization of FHL2, a novel LIM domain protein preferentially expressed in human heart. Gene 210: 345–350

Chrivia JC, Kwok RP, Lamb N, Hagiwara M, Montminy MR, Goodman RH (1993) Phosphorylated CREB binds specifically to the nuclear protein CBP. Nature 365: 855–859

Cohen P (1997) The search for physiological substrates of MAP and SAP kinases in mammalian cells. Trends Cell Biol 7: 353–361

Courey AJ, Tjian R (1988) Analysis of Sp1 in vivo reveals multiple transcriptional domains, including a novel glutamine-rich activation motif. Cell 55: 887–898

Dawid IB, Breen JJ, Toyama R (1998) LIM domains: multiple roles as adapters and functional modifiers in protein interactions. Trends Genet 14: 156–162

De Cesare D, Fimia GM, Sassone-Corsi P (1999) Signaling routes to CREM and CREB: plasticity in transcriptional activation. Trends Biochem Sci 24: 281–285

De Cesare D, Jacquot S, Hanauer A, Sassone-Corsi P (1998) Rsk-2 activity is necessary for epidermal growth factor-induced phosphorylation of CREB protein and transcription of c-*fos* gene. Proc Natl Acad Sci USA 95: 12202–12207

de Groot R, Ballou LM, Sassone-Corsi P (1994) Positive regulation of the cAMP-responsive activator CREM by the p70 S6 kinase: an alternative route to mitogen-induced gene expression. Cell 79: 81–91

de Groot RP, den HJ, Vandenheede JR, Goris J, Sassone-Corsi P (1993) Multiple and cooperative phosphorylation events regulate the CREM activator function. EMBO J 12: 3903–3911

Deak M, Clifton AD, Lucocq LM, Alessi DR (1998) Mitogen- and stress-activated protein kinase-1 (MSK1) is directly activated by MAPK and SAPK2/p38, and may mediate activation of CREB. EMBO J 17: 4426–41

Delmas V, van dHF, Mellstrom B, Jegou B, Sassone-Corsi P (1993) Induction of CREM activator proteins in spermatids: down-stream targets and implications for haploid germ cell differentiation. Mol Endocrinol 7: 1502–1514

Eckner R, Ewen ME, Newsome D, Gerdes M, De CJ, Lawrence JB, Livingston DM (1994) Molecular cloning and functional analysis of the adenovirus E1A-associated 300-kD protein (p300) reveals a protein with properties of a transcriptional adaptor. Genes Dev 8: 869–884

Ferreri K, Gill G, Montminy M (1994) The cAMP-regulated transcription factor CREB interacts with a component of the TFIID complex. Proc Natl Acad Sci USA 91: 1210–1213

Fimia GM, De Cesare D, Sassone-Corsi P (1998) Mechanisms of activation by CREB and CREM: phosphorylation, CBP, and a novel coactivator, ACT. Cold Spring Harb Symp Quant Biol 63: 631–642

Fimia GM, De Cesare D, Sassone-Corsi P (1999) CBP-independent activation of CREM and CREB by the LIM-only protein ACT. Nature 398: 165–169

Foulkes NS, Borjigin J, Snyder SH, Sassone-Corsi P (1997) Rhythmic transcription: the molecular basis of circadian melatonin synthesis. Trends Neurosci 20: 487–492

Foulkes NS, Borrelli E, Sassone-Corsi P (1991) CREM gene: use of alternative DNA-binding domains generates multiple antagonists of cAMP-induced transcription. Cell 64: 739–749

Foulkes NS, Mellstrom B, Benusiglio E, Sassone-Corsi P (1992) Developmental switch of CREM function during spermatogenesis: from antagonist to activator. Nature 355: 80–84

Foulkes NS, Schlotter F, Pevet P, Sassone-Corsi P (1993) Pituitary hormone FSH directs the CREM functional switch during spermatogenesis. Nature 362: 264–267

Genini M, Schwalbe P, Scholl FA, Remppis A, Mattei MG, Schafer BW (1997) Subtractive cloning and characterization of DRAL, a novel LIM-domain protein down-regulated in rhabdomyosarcoma. DNA Cell Biol 16: 433–442

Ginty DD, Bonni A, Greenberg ME (1994) Nerve growth factor activates a Ras-dependent protein kinase that stimulates c-fos transcription via phosphorylation of CREB. Cell 77: 713–725

Gonzalez GA, Montminy MR (1989) Cyclic AMP stimulates somatostatin gene transcription by phosphorylation of CREB at serine 133. Cell 59: 675–680

Hai TW, Liu F, Coukos WJ, Green MR (1989) Transcription factor ATF cDNA clones: an extensive family of leucine zipper proteins able to selectively form DNA-binding heterodimers [published erratum appears in Genes Dev 1990 Apr; 4(4):682]. Genes Dev 3: 2083–2090

Iordanov M, Bender K, Ade T, Schmid W, Sachsenmaier C, Engel K, Gaestel M, Rahmsdorf HJ, Herrlich P (1997) CREB is activated by UVC through a p38/HOG-1-dependent protein kinase. EMBO J 16: 1009–1022

Jégou B (1993) The Sertoli-germ cell communication network in mammals. Int Rev Cytol 147: 25–96

Kistler MK, Sassone-Corsi P, Kistler WS (1994) Identification of a functional cyclic adenosine 3',5'-monophosphate response element in the 5'-flanking region of the gene for transition protein 1 (TP1), a basic chromosomal protein of mammalian spermatids. Biol Reprod 51: 1322–1329

Kouzarides T (1999) Histone acetylases and deacetylases in cell proliferation. Curr Opin Genet Dev 9: 40–48

Kwok RP, Lundblad JR, Chrivia JC, Richards JP, Bachinger HP, Brennan RG,

Roberts SG, Green MR, Goodman RH (1994) Nuclear protein CBP is a coactivator for the transcription factor CREB. Nature 370: 223–226

Lalli E, Sassone-Corsi P (1994) Signal transduction and gene regulation: the nuclear response to cAMP. J Biol Chem 269: 17359–17362

Lamas M, Sassone-Corsi P (1997) The dynamics of the transcriptional response to cyclic adenosine 3',5'-monophosphate: recurrent inducibility and refractory phase. Mol Endocrinol 11: 1415–1424

Laoide BM, Foulkes NS, Schlotter F, Sassone-Corsi P (1993) The functional versatility of CREM is determined by its modular structure. EMBO J 12: 1179–1191

McKnight GS, Clegg CH, Uhler MD, Chrivia JC, Cadd GG, Correll LA, Otten AD (1988) Analysis of the cAMP-dependent protein kinase system using molecular genetic approaches. Recent Prog Horm Res 44: 307–335

Molina CA, Foulkes NS, Lalli E, Sassone-Corsi P (1993) Inducibility and negative autoregulation of CREM: an alternative promoter directs the expression of ICER, an early response repressor. Cell 75: 875–886

Montminy M (1997) Transcriptional regulation by cyclic AMP. Annu Rev Biochem 66: 807–822

Morgan MJ, Madgwick AJA (1996) Slim defines a novel family of LIM-proteins expressed in skeletal muscle. Biochem Biophys Res Commun 225: 632–638

Morgan MJ, Madgwick AJA (1999) The fourth member of the FHL family of LIM proteins is expressed exclusively in the testis. Biochem Biophys Res Commun 255: 251–255

Nakajima T, Uchida C, Anderson SF, Lee CG, Hurwitz J, Parvin JD, Montminy M (1997) RNA helicase A mediates association of CBP with RNA polymerase II. Cell 90: 1107–1112

Nantel F, Monaco L, Foulkes NS, Masquilier D, Le MM, Henriksen K, Dierich A, Parvinen M, Sassone-Corsi P (1996) Spermiogenesis deficiency and germ-cell apoptosis in CREM-mutant mice. Nature 380: 159–162

Radhakrishnan I, Perez-Alvarado GC, Parker D, Dyson HJ, Montminy MR, Wright PE (1997) Solution structure of the KIX domain of CBP bound to the transactivation domain of CREB: a model for activator:coactivator interactions. Cell 91: 741–752

Rehfuss RP, Walton KM, Loriaux MM, Goodman RH (1991) The cAMP-regulated enhancer-binding protein ATF-1 activates transcription in response to cAMP-dependent protein kinase A. J Biol Chem 266: 18431–18434

Roesler WJ, Vandenbark GR, Hanson RW (1988) Cyclic AMP and the induction of eukaryotic gene transcription. J Biol Chem 263: 9063–9066

Rozman D, Fink M, Fimia GM, Sassone-Corsi P, Waterman MR (1999) cAMP/CREM-dependent regulation of cholesterogenic lanosterol 14-alpha demethylase (CYP51) in spermatids. Mol Endocrinol 13: 1951–1962.

Sassone-Corsi P (1995) Transcription factors responsive to cAMP. Annu Rev Cell Dev Biol 11: 355–377

Sassone-Corsi P (1997) Transcriptional checkpoints determining the fate of male germ cells. Cell 88: 163–166

Sassone-Corsi P (1998) CREM: a master-switch governing male germ cell differentiation and apoptotis. Sem Cell Dev Biol 9: 475–482

Sassone-Corsi P, Visvader J, Ferland L, Mellon PL, Verma IM (1988) Induction of proto-oncogene fos transcription through the adenylate cyclase pathway: characterization of a cAMP-responsive element. Genes Dev 2: 1529–1538

Schmeichel KL, Beckerle MC (1994) The LIM domain is a modular protein-binding interface. Cell 79: 211–219

Sheng M, Thompson MA, Greenberg ME (1991) CREB: a Ca2+-regulated transcription factor phosphorylated by calmodulin-dependent kinases. Science 252: 1427–1430

Shikama N, Lyon J, La Thangue NB (1997) The p300/CBP family: integrating signals with transcription factors and chromatin. Trends Cell Biol 7: 230–236

Silva AJ, Kogan JH, Frankland PW, Kida S (1998) CREB and memory. Annu Rev Neurosci 21: 127–148

Struthers RS, Vale WW, Arias C, Sawchenko PE, Montminy MR (1991) Somatotroph hypoplasia and dwarfism in transgenic mice expressing a non-phosphorylatable CREB mutant. Nature 350: 622–624

Sun Z, Sassone-Corsi P, Means AR (1995) Calspermin gene transcription is regulated by two cyclic AMP response elements contained in an alternative promoter in the calmodulin kinase IV gene. Mol Cell Biol 15: 561–571

Tan Y, Rouse J, Zhang A, Cariati S, Cohen P, Comb MJ (1996) FGF and stress regulate CREB and ATF-1 via a pathway involving p38 MAP kinase and MAPKAP kinase-2. EMBO J 15: 4629–4642

Taniguchi Y, Furukawa T, Tun T, Han H, Honjo T (1998) LIM protein KyoT2 negatively regulates transcription by association with the RBP-J DNA-binding protein. Mol Cell Biol 18: 644–654

Williams T, Admon A, Luscher B, Tjian R (1988) Cloning and expression of AP-2, a cell-type-specific transcription factor that activates inducible enhancer elements. Genes Dev 2: 1557–1569

Xing J, Ginty DD, Greenberg ME (1996) Coupling of the RAS-MAPK pathway to gene activation by RSK2, a growth factor-regulated CREB kinase. Science 273: 959–963

Zhou Y, Sun Z, Means AR, Sassone CP, Bernstein KE (1996) cAMP-response element modulator tau is a positive regulator of testis angiotensin converting enzyme transcription. Proc Natl Acad Sci USA 93: 12262–12266

9 Apoptosis – Searching for the Central Executioner

E. Daugas, G. Kroemer

9.1 Introduction: the Notion of the Central Executioner 129
9.2 Caspases: Central to the Extrinsic Pathway 131
9.3 Caspase-Independent Cell Death and Apoptosis 133
9.4 Mitochondria: Central to the Intrinsic Pathway 134
References 137

Abbreviations. BA: bongkrekic acid, CsA: cyclosporin A, $\Delta\Psi_m$: mitochondrial transmembrane potential, PTPC: permeability transition pore complex

9.1 Introduction: the Notion of the Central Executioner

Apoptosis may be defined as a regulated lethal process in which the cell activates catabolic processes which, within the limits of a near-to-intact plasma membrane, lead to a stereotyped ensemble of biochemical and morphological alterations. Such alterations include a reduction in cell size, a condensation of chromatin, and changes in the physicochemistry of the plasma membrane facilitating the recognition and heterophagic removal of the apoptotic cell by adjacent normal cells. The most striking morphological change in apoptotic cells concerns the nucleus which invariably exhibits chromatin condensation, mostly associated with enzymatic degradation of nuclear DNA. However, chromatin condensation is a sign of apoptosis rather than a mechanism leading to cell death, since non-nuclear apoptosis-associated alterations can be induced in anucleate cells (cytoplasts), as this has been shown in 1994 (Jacobson et al. 1994, Schulze-Osthoff et al. 1994). This observation invalidated the earlier hypothesis that DNases would actively participate in the "decision to die" (McConkey et al. 1989) and led to the postulation of a cytoplasmic (non-nuclear) effector or "central executioner" that would participate in life/death decision making and would be influenced by endogenous control mechanisms (Jacobson et al. 1994, Schulze-Osthoff et al. 1994). From 1995, most investigators in the field felt that apoptosis would depend on the activation of a class of cysteine proteases cleaving after Asp residues ("caspases"), based on the observation that the inhibition of such enzymes prevents most of the apoptotic morphological and biochemical alterations and can retard the loss of cell viability (Martin and Green 1995). The hypothesis that caspases would constitute the (only) "central executioner", however, had to be abandoned in face of the facts that (i) morphological signs of apoptosis may occur without caspase activation (see below), (ii) that pharmacological inhibition of caspases mostly fails to prevent cell death, as defined by a loss of clonogenic potential (Green and Kroemer 1998), and (iii) that caspase activation may occur without cell death (Alam et al. 1999, Kennedy et al. 1999). These findings thus invalidated the idea that the cytoplasmic central execution could be identical with a cascade of proteolytic enzymes. Rather, it appears today that caspases play a dual role in the apoptotic process. On the one hand, caspases may function as signal transducing molecules when cell death is induced by the so-called "extrinsic pathway" and thus connect so-called death receptors (e.g. Fas/Apo-1/CD95, TNF-R) to the apoptotic machinery. When caspases act as signal transducers, their activation is required for apoptosis induction, and their inhibition thus prevents cell death. On the other hand, in most models of apoptosis induction (which involves the "intrinsic" pathway) caspases only participate in the apoptotic degradation phase, after activation of the central executioner (who seals the cell's fate). To use a drastic expression, the contribution of caspases would then be reduced to their participation in the putrefaction of the dead body, after death has ensued in a caspase-independent fashion.

What is then the central executioner, beyond the putative contribution of caspases? Based on theoretical considerations, seven criteria should be fulfilled by the hypothetical central executioner (Kroemer 1997a, Penninger and Kroemer 1998). First, the central executioner should become activated during the effector stage of apoptosis, at the point-of-no-return, *before* the manifestations of the apoptotic degradation phase (nuclear chromatin condensation and fragmentation, phosphatidylserine exposure, advanced proteolysis of vital proteins etc.) become manifest (chronological criterion). Second, the central executioner should constitute an essential feature of the apoptotic process and should be undissociable from naturally occurring apoptosis (functional criterion). Third, the central executioner should be triggered by many different apoptosis induction protocols, independently from the pro-apoptotic trigger – receptor-mediated signals or damage – thus allowing for the convergence of different stimulus-dependent signal transduction pathways into one single pathway (criterion of convergence). Fourth, the central executioner should be capable of coordinating the different manifestations of apoptosis at the levels of the nucleus, the cytoplasm, and the plasma membrane (criterion of coordination). Fifth, given the fact that all cell types can de driven into apoptosis, even in the presence of protein synthesis inhibitors, the central executor should be pre-existing in all cells (criterion of universality). Sixth, since even cancer cells fail to develop a complete resistance to apoptosis induction, the central executioner or its compounds must have some function(s) that is/are essential for normal cell survival (criterion of vitality). Seventh, since cell death is an all-or-nothing phenomenon, the central executioner should function as a switch that is either off or on (criterion of the switch). The present review will summarize evidence indicating that mitochondria play a major role in the apoptotic process, at least in the caspase-independent intrinsic pathway.

9.2 Caspases: Central to the Extrinsic Pathway

A series of so-called death receptors can directly trigger direct caspase upon ligation. Thus, the first death receptor to be defined in molecular terms, Fas/Apo-1/CD95, which is expressed in the plasma membrane, possesses a so-called death domain (DD) in its cytoplasmic region. This

DD interacts with another DD of a cytosolic adapter molecule, FADD. In addition to DD, FADD contains a death effector domain (DED), which interacts with the DED of the enzymatically inactive precursor of caspase-8 (pro-caspase-8). Upon occupancy of CD95 by its ligand (CD95L), CD95 trimerizes and recruits FADD as well as pro-caspase-8, which proteolytically auto-activates within the so-called DISC (death-induced signaling complex) (Peter and Krammer 1998, Budijardjo et al. 1999). Active caspase-8 then is released from the DISC and digest other pro-caspases (e.g. pro-caspase-3) causing their activation. Caspase-3 then cleaves several vital proteins and leads to the activation of caspase-activated DNase (CAD), an enzyme which digests chromatin to short fragments (mono- or oligomers of 200 bp). Alternatively, caspase-8 cleaves the pro-apoptotic Bcl-2 family protein Bid, leading to its N-terminal truncation (tBid). tBid translocates from the cytosol to mitochondrial membranes and triggers the release of cytochrome *c* (normally confined to the mitochondrial intermembrane space) into the extra-mitochondrial compartment. Cytochrome *c* then interacts with cytosolic Apaf-1, resulting in its oligomerization and the recruitment of pro-caspase-9, which becomes activated within the so-called apoptosome (Budijardjo et al. 1999, Salvesen and Dixit 1999). Active caspase-9 can also activate pro-caspase-3. The relative abundance of proteins recruited into the DISC (and perhaps hitherto unknown qualitative differences) determine whether pro-caspase-8 activation leads to direct activation of pro-caspase-3 (which is the case in so-called type-1 cells) or whether pro-caspase-3 activation requires the contribution of Bid/tBid and mitochondrial factors (the case of so-called type-2 cells) (Scaffidi et al. 1998, Yin et al. 1999). Only in this latter case, can Bcl-2 (which prevents tBid-induced mitochondrial cytochrome *c* release) prevent cell death.

The Fas/Apo-1/CD95-triggered signaling cascade constitutes the paradigm of the "extrinsic" pathway culminating into cell death. Similar pathways have been delineated for the tumor necrosis factor receptor (TNF-R) and other death receptors including DR3 (Budijardjo et al. 1999). Inhibition of caspases by specific gene products (e.g. CrmA or FLIP, which inhibit caspase-8 activation) or modified peptide pseudosubstrates (e.g. Z-VAD.fmk) can prevent signaling via Fas/Apo-1/CD95, thus preventing the induction of cell death (Peter and Krammer 1998), as well as the transmission of Fas/Apo-1/CD95-triggered mitogenic signals (Alam et al. 1999, Kennedy et al. 1999). In this context, caspase-8 thus may be considered as an obligate constituent of a pro-apoptotic signal transduction pathway. By analogy, other caspases may also be involved in lethal signal transduction. Thus, the knock-out of the caspase-3 gene causes defective death of post-mitotic neurons, though does not affect the death of other cell types including thymocytes (Colussi and Kumar 1999).

9.3 Caspase-Independent Cell Death and Apoptosis

Pharmacological inhibition of caspases can prevent the acquisition of certain hallmarks of the apoptotic process, in particular the oligonucleosomal DNA fragmentation mediated by CAD. Thus, irrespective of the model that is investigated, caspase inhibition prevents the digestion of nuclear DNA into mono- and oligomers of ~200 bp. However, in most models of apoptosis, with the exception of those triggered via the extrinsic pathway (in which caspases function as obligate signal transducers), inhibition of caspases does not prevent cell death as such and rather results into delayed cytolysis in which some features of apoptosis fail to become apparent (Table 1). This applies to chemother-

Table 1. Models of apoptotic cell death in which caspase inhibition fails to confer cytoprotection

Model of cell death	Examples
Developmental cell death	Interdigital cell death
Hormone-induced cell death	Glucocorticoid receptor occupancy in thymocytes
Ligation of cell surface receptors	CD2, CD4, CD45, CD47, CD99, class II
Death of T cell targets	Perforin + granzyme – induced cytolysis
Chemotherapy	Etoposide, camptothecin, CD437
Pro-apoptotic Bcl-2-like proteins	Bax, Bak
Oncogenes	c-Myc, PML, Ras
Pro-apoptotic second messengers	Ceramide, ganglioside GD3, reactive oxygespecies
Viral proteins	E4orf4 of adenovirus, Vpr of HIV-1
Overexpression of apoptotic effectors	FADD, CED4, AIF

apy-induced cell death (e.g. that induced by etoposide or camptothecin), developmental cell death (e.g. interdigital cell death), death induced by some pro-apoptotic signal transducers (e.g. ceramide and ganglioside GD3), death induced by overexpression of pro-apoptotic Bcl-2 family proteins (Bax, Bak), death induced by some viral products (e.g. E4orf4 of adenovirus, Vpr of HIV-1), and death induced via a variety of intracellular receptors (e.g. the glucocorticoid receptor) or cell surface receptors (CD2, CD4, CD45, CD99, CXCR4 etc.) (Deas et al. 1998, Chautan et al. 1999, Kitanaka and Kuchino 1999, Jacotot et al. 2000). In several instances, hallmarks of apoptosis such as chromatin condensation can be induced without that caspase activation occurs at all. Examples for this death modality include the overexpression of the oncogene PML (Quignon et al. 1998) or cross-linking of CXCR4 (Berndt et al. 1998). In most cases, caspase activation normally does occur, but its inhibition does not prevent the loss of clonogenic survival and cell viability. As a result, the caspase activation cascade cannot be considered as being a major part of the central executioner. Caspases do not fulfill all criteria enumerated in the introduction of this article. Although their activation suffices to cause apoptosis and cell death, they are not (always) required for cell death induction.

9.4 Mitochondria: Central to the Intrinsic Pathway

The "intrinsic" pathway may be defined by antinomy with the strictly caspase-dependent, death receptor-mediated extrinsic pathway (see above). Irrespective of the apoptosis trigger, mitochondrial membrane permeabilization occurs at an early stage of apoptosis, before chromatin condensation and DNA fragmentation become visible. Signs of membrane permeabilization include release of intermembrane proteins from mitochondria, physical disruption of the outer membrane while parts of the inner membrane herniate through the gaps, loss of inner membrane christae, swelling of the mitochondrial matrix, and/or loss of the inner transmembrane potential (Kroemer et al. 1997, Kroemer et al. 1998). Although the exact molecular mechanisms underlying these phenomena may be diverse, the signs of mitochondrial membrane permeabilization appear to be a near-to-constant, caspase-independent feature of cell death triggered via the intrinsic pathway. Is the mitochondrion then part of the central executioner? We will briefly reconsider here the criteria established in the introduction of this paper.

Chronological Criterion. Mitochondrial membrane alterations affecting either the inner and/or outer mitochondrial membranes appear to be a constant feature of early apoptosis (Kroemer et al. 1997, Kroemer et al. 1998). Failure to detect such alterations is mostly due to the use of inadequate methodologies (Bernardi et al. 1999).

Functional Criterion. Inhibition of mitochondrial membrane permeabilization by anti-apoptotic members of the Bcl-2 family (Bcl-2, Bcl-X_L), which are locally present in mitochondrial membranes suffices to retard and/or to prevent cell death, whenever their local presence succeeds in avoiding membrane permeabilization (Susin et al. 1996, Kroemer 1997b, Kluck et al. 1997). Induction of membrane permeabilization by agents that directly act on mitochondrial membranes suffices to kill cells (Marchetti et al. 1996). As a result, mitochondrial membrane permeabilization is both necessary for cell death to occur (at least in the intrinsic pathway and, in the extrinsic pathway, in type 2 cells) and sufficient to kill cells.

Criterion of Convergence. A growing body of lethal molecules directly act on mitochondria (Table 2). This has been clearly established for pro-apoptotic members of the Bcl-2 family, a number of non-protein pro-apoptotic second messengers, several viral proteins, as well as cytotoxic compounds used in experimental cancer therapy. As a result, very different apoptosis inducers converge on mitochondria.

Table 2. Molecules directly acting on mitochondria to trigger cell death

Class of molecules	Examples
Pro-apoptotic members of the Bcl-2 family	Bax, Bak, Bid, Bik, Bim
Pro-apoptotic second messengers	Ca^{2+}, ganglioside GD3, ROS, NO
Viral effectors	Vpr of HIV-1
	Tat of HIV-1
	p13II of HTLV-1
	HBV protein X of hepatitis B virus
Xenobiotics	Lonidamine, CD437, arsenic oxide, diamide, betulinic acid

Table 3. Lethal proteins released from mitochondria upon induction of apoptosis

Class of molecules	Examples
Caspase activators	cytochrome *c*
	10 kd heat shock protein
Caspases	caspases 2, 7, 9
Catabolic enzymes	arginase 1
	glycine cleavage system h protein
	lysozyme homologue AT 2
	soluble epoxide hydrolase
	sulfite oxidase

Criterion of Coordination. Inhibition of mitochondrial membrane permeabilization by overexpression of Bcl-2 or by pharmacological effectors specifically acting on mitochondrial proteins prevents all signs of apoptosis, at the levels of the plasma membrane, cytoplasm and nucleus, indicating that mitochondrial permeabilization constitutes a critical coordinating step of the apoptotic porches (Marchetti et al. 1996)s. How is this coordination established? On the one hand, mitochondria release a whole series of proteins that trigger catabolic reactions (Table 3). This applies to cytochrome *c* (see above), a critical caspase activator (Liu et al. 1996), as well as to some pro-caspases which are specifically enriched in the mitochondrial intermembrane space (Susin et al. 1999a). In addition, mitochondria release AIF (apoptosis inducing factor), a flavoenzyme which translocates to the nucleus and may account for large-scale DNA fragmentation to ~50 kbp fragments and peripheral chromatin condensation (Susin et al. 1999b, Lorenzo et al. 1999). Mass spectroscopy has recently lead to the identification of further catabolic enzymes which can be released upon mitochondrial membrane permeabilization (Patterson et al. 2000). Moreover, loss of mitochondrial function ultimately results in the depletion of ATP and in a major imbalance in redox metabolism which eventually will lead to cell death.

Criteria of Universality and Vitality. It appears obvious that mitochondria are contained in all cells and that their presence is vital for cell survival.

Criterion of the Switch. Permeabilization of mitochondrial membranes has multiple consequences (activation of downstream caspases, increase in cytosolic free Ca2+, overproduction of reactive oxygen species), which themselves trigger mitochondrial membrane permeabilization (Kroemer 1997a, Kroemer et al. 1997, Kroemer et al. 1998, Penninger and Kroemer 1998). This positive feed forward loop determines that, beyond a critical threshold level, mitochondrial membrane permeabilization is a self-amplifying process which acts in an all-or-nothing fashion.

In conclusion, mitochondria are part of the central executioner of cell death.

Acknowledgements. The authors' own work has been supported by a special grant from the Ligue Nationale contre le Cancer (to G.K.). E.D. is sponsored by CNAMTS/APHP.

References

Alam A, Cohen LY, Aouad S, Sekaly RP (1999) Early activation of caspases during T lymphocyte stimulation results in selective substrate cleavage in nonapoptotic cells J. Exp. Med. 190: 1879–1890

Bernardi P, Scorrano L, Colonna R, Petronilli V, Di Lisa F (1999) Mitochondria and cell death – Mechanistic aspects and methodological issues Eur. J. Biochem. 264: 687–701

Berndt C, Möpps B, Angermüller S, Gierschik P, Krammer PH (1998) CXCR4 and CD4 mediate a rapid CD95-independent cell death in CD4+ cells Proc. Natl. Acad. Sci. USA 95: 12556–12561

Budijardjo I, Oliver H, Lutter M, Luo X, Wang X (1999) Biochemical pathways of caspase activation during apoptosis Annu. Rev. Cell Dev. Biol. 15: 269–290

Chautan M, Chazal G, Cecconi F, Gruss P, Golstein P (1999) Interdigital cell death can occur through a necrotic and caspase-independent pathway Curr. Biol. 9: 967–970

Colussi PA, Kumar S (1999) Targeted disruption of caspase genes in mice: What they tell us about the functions of individual caspases in apoptosis Immunol. Cell Biol. 77: 58–63

Deas O, Dumont C, MacFarlane M, Rouleau M, Hebib C, Harper F, Hirsch F, Charpentier B, Cohen GM, Senik A (1998) Caspase-independent cell death

induced by anti-CD2 or staurosporine in activated human peripheral T lymphocytes J. Immunol. 161: 3375–3383
Green DR, Kroemer G (1998) The central executioner of apoptosis: mitochondria or caspases? Trends Cell Biol. 8: 267–271
Jacobson MD, Burne JF, Raff MC (1994) Programmed cell death and Bcl-2 protection in the absence of a nucleus EMBO J. 13: 1899–1910
Jacotot E, Ravagnan L, Loeffler M, Ferri KF, Vieira HLA, Zamzami N, Costantini P, Druillennec S, Hoebeke J, Brian JP, Irinopoulos T, Daugas E, Susin SA, Cointe D, Xie ZH, Reed JC, Roques BP, Kroemer G (2000) The HIV-1 viral protein R induces apoptosis via a direct effect on the mitochondrial permeability transition pore J. Exp. Med. 191: 33–45
Kennedy NJ, Kataoka T, Tschopp J, Budd RC (1999) Caspase activation is required for T cell proliferation J. Exp. Med. 190: 1891–1895
Kitanaka C, Kuchino Y (1999) Caspase-independent programmed cell death with necrotic morphology Cell Death Differ. 6: 508–515
Kluck RM, Bossy-Wetzel E, Green DR, Newmeyer DD (1997) The release of cytochrome c from mitochondria: a primary site for Bcl-2 regulation of apoptosis Science 275: 1132–1136
Kroemer G (1997a) Mitochondrial implication in apoptosis. Towards an endosymbiotic hypothesis of apoptosis evolution Cell Death Differentiation 4: 443–456
Kroemer G (1997b) The proto-oncogene Bcl-2 and its role in regulating apoptosis Nature Medicine 3: 614–620
Kroemer G, Dallaporta B, Resche-Rigon M (1998) The mitochondrial death/life regulator in apoptosis and necrosis Annu. Rev. Physiol. 60: 619–642
Kroemer G, Zamzami N, Susin SA (1997) Mitochondrial control of apoptosis Immunol. Today 18: 44–51
Liu XS, Kim CN, Yang J, Jemmerson R, Wang X (1996) Induction of apoptotic program in cell-free extracts: requirement for dATP and cytochrome C Cell 86: 147–157
Lorenzo HK, Susin SA, Penninger J, Kroemer G (1999) Apoptosis inducing factor (AIF): a phylogenetically old, caspase-independent effector of cell death Cell Death Differ. 6: 516–524
Marchetti P, Castedo M, Susin SA, Zamzami N, Hirsch T, Haeffner A, Hirsch F, Geuskens M, Kroemer G (1996) Mitochondrial permeability transition is a central coordinating event of apoptosis J. Exp. Med. 184: 1155–1160
Martin SJ, Green DR (1995) Protease activation during apoptosis: death by a thousand cuts? Cell 82: 349–352
McConkey DJ, Hartzell P, Nicotera P, Orrenius S (1989) Calcium-activated DNA fragmentation kills immature thymocytes FASEB J. 3: 1843–1849

Patterson S, Spahr CS, Daugas E, Susin SA, Irinopoulos T, Koehler C, Kroemer G (2000) Mass spectrometric identification of proteins released from mitochondria undergoing permeability transition Cell Death Differ. in press:

Penninger JM, Kroemer G (1998) Molecular and cellular mechanisms of T lymphocyte apoptosis Adv. Immunol. 68: 51–144

Peter ME, Krammer PH (1998) Mechanisms of CD95 (APO-1/Fas)-mediated apoptosis Curr. Op. Immunol. 10: 545–551

Quignon F, DeBels F, Koken M, Feunteun J, Ameisen JC, de Thé H (1998) PML induces a novel caspase-independent death process Nat. Gen. 20: 259–265

Salvesen GS, Dixit VM (1999) Caspase activation: the induced-proximity model Proc. Natl. Acad. Sci. USA 96: 10964–10967

Scaffidi C, Fulda S, Srinivasan A, Friesen C, Li F, Tomaselli KJ, Debatin K-M, Krammer PH, Peter ME (1998) Two CD95 (APO-1/Fas) signaling pathways EMBO J. 17: 1675–1687

Schulze-Osthoff K, Walczak H, Droge W, Krammer PH (1994) Cell nucleus and DNA fragmentation are not required for apoptosis J. Cell Biol. 127: 15–20

Susin SA, Lorenzo HK, Zamzami N, Marzo I, Larochette N, Alzari PM, Kroemer G (1999a) Mitochondrial release of caspases-2 and –9 during the apoptotic process J. Exp. Med. 189: 381–394

Susin SA, Lorenzo HK, Zamzami N, Marzo I, Snow BE, Brothers GM, Mangion J, Jacotot E, Costantini P, Loeffler M, Larochette N, Goodlett DR, Aebersold R, Siderovski DP, Penninger JM, Kroemer G (1999b) Molecular characterization of mitochondrial apoptosis-inducing factor Nature 397: 441–446

Susin SA, Zamzami N, Castedo M, Hirsch T, Marchetti P, Macho A, Daugas E, Geuskens M, Kroemer G (1996) Bcl-2 inhibits the mitochondrial release of an apoptogenic protease J. Exp. Med. 184: 1331–1342

Yin X-M, Wang K, Gross A, Zhao Y, Zinkel S, Klocke B, Rothe KA, Korsmeyer SJ (1999) Bid-deficient mice are resistant to Fas-induced hepatocellular apoptosis Nature 400: 886–89

10 Germ Cell Apoptosis

M. E. Embree, K. Boekelheide

10.1 Introduction . . . 141
10.2 Background . . . 142
10.3 p53/FasL Double Deficient Mouse . . . 151
10.4. Summary and Integration . . . 158
References . . . 159

10.1 Introduction

This presentation begins with a review of the major systems involved in the control of germ cell apoptosis, setting the stage for a discussion of ongoing experimental work from our laboratory. Because of the now vast literature on apoptosis in general and germ cell apoptosis specifically, we do not attempt to be comprehensive. Rather, our goal is to provide sufficient information to lead the reader to pertinent references, and to make the discussion of our ongoing experiments comprehensible.

The description of important apoptotic systems is limited to p53, the Fas system, *bcl-2* family members, and the caspases. These are some of the major known participants in germ cell apoptosis; however, it should be recognized by the reader that additional apoptotic systems, both those already discovered and yet to be discovered, are likely to play significant roles in the process of programmed germ cell death. Even though we know a lot about some specific aspects of the process of apoptosis in the testis, it is clear that we have only limited knowledge of those features of cellular regulation involved in integrating the various signaling and execution pathways.

10.2 Background

This Background section is divided into three parts. In an initial overview section, general features of testicular germ cell apoptosis are discussed. This is followed by a presentation of four important apoptotic systems – p53, the Fas system, *bcl-2* family members, and the caspases. A final section examines different experimental approaches which have been used to dissect the role of apoptosis in testicular homeostasis.

10.2.1 Overview

Early work in testis and spermatogenesis focused on the role of growth and survival factors in the development and maintenance of "normal" germ cell function. However, it has become increasingly evident that in addition to germ cell proliferation and survival, cell death is important to testicular homeostasis. Apoptosis occurs at a baseline level in the adult testis, such that 25–75% of the expected germ cell yield is lost during spermatogenesis (Oakberg 1956, Huckins 1978, De Rooij and Lok 1987), mostly through apoptosis of spermatogonia and spermatocytes (Allan et al. 1992, Bartke 1995, Billig et al. 1995, and Blanco-Rodriguez and Martinez-Garcia 1996). In addition, there are several stages during development when waves of apoptosis occur in germ cells (Coucouvanis et al 1993, Wang et al. 1998a).

Our current understanding of apoptosis comes from two major types of observations: the characterization of the molecular events involved in apoptosis and the description of morphological changes associated with apoptotic cells. Caenorhabditis elegans provided the first molecular clues to the identification of a series of genes – the *ced* genes and their mammalian homologs – involved in the intracellular cascade of genetically programmed cell death. The strategy of using more experimentally and genetically tractable model systems to identify important apoptotic pathways common with mammalian systems has been very successful and speaks to the highly conserved nature of these pathways. The typical morphology of an apoptotic cell – chromatin condensation, cell shrinkage, and cytoplasmic blebbing – reflects the systematic disassembly of the key components of the cell by enzymes activated in the apoptotic cascade. Of note, although many of the generally known genes involved in apoptosis are present in germ cells in the testis, the morphology of dying germ cells often does not conform to the traditional description of apoptotic cells. Spermatocytes undergoing apoptosis, in particular, are often described as having a fragmented nucleus more characteristic of necrosis, rather than the chromatin condensation typical of apoptotic cells. Yet, these dying spermatocytes are assumed to be undergoing apoptosis because of the presence of DNA ladders in agarose gels, positive staining for DNA end-labeling, the timing of the cell death relative to cell cycle, and up-regulation of apoptotic genes (Blanco-Rodriguez and Martinez-Garcia 1996, Brinkworth et al 1995, Hikim et al. 1997, Boekekheide et al. 1998). The difference in morphologic appearance of apoptotic germ cells may be a reflection of alternate molecular mechanisms involved in germ cell apoptosis or the special biology of these cells.

Unlike necrosis – in which the cell dies without activation of any machinery – apoptosis is an active process with the initiating signal coming either through extracellular cell surface receptors or from intracellular sensors. These signals lead to activation of a cascade of enzymes which are responsible for the characteristic morphology of apoptotic cells: DNA is cleaved into distinct fragments (multiples of 180–200 b.p.) which are evident as a DNA ladder when DNA is submitted to gel electrophoresis; proteins in the cytoskeleton are disassembled and/or crosslinked; and the lipid membrane at the cell surface changes composition (such as the flipping of phosphatidyl serine in the lipid bilayer). The apoptotic process can be triggered by the activation of death receptors, such as Fas, or by the induction of intracellular "stress" sensors. The cascade can also be turned on by the absence of a needed trophic signal which normally promotes cell survival. In other words, apoptotic cell death may be evoked because of neglect. Given the known important roles of Sertoli and Leydig cells in supporting testicular germ cells, it is likely that intercellular communication through a number of paracrine regulators mediates the germ cell apoptotic process.

10.2.2 Apoptotic Systems

10.2.2.1 p53

p53, most commonly known as a tumor suppresser, has a dual role as a cell cycle check and apoptotic inducer; which role it plays depends on the cell type, the state of the cell and the initiating signal. It is not surprising, then, that p53 interacts with many different molecules in very distinct ways. It can act as a transcription factor (for example, for *bax*) (Miyashita 1995), as a transcriptional repressor (for example, of *bcl-2*) (Selvakumaran et al. 1994) and as an inhibitor of protein activity (such as for the helicase activity of TFIIH protein complex) (Wang et al. 1995 and Wang et al. 1996). Because of the complex nature of p53 - the wealth of signals which activate p53 and the number of downstream effecters that p53 can influence – it is often difficult to tease out the precise role that p53 plays in a specific response.

The complex biology of p53 is closely associated with its structure (for reviews, see Agarwal 1998 and Bellamy 1997). The C-terminus of the protein contains the sequence-specific DNA binding domain. Composed primarily of basic residues, this domain is subject to modification by acetylation, phosphorylation, *O*-glycosylation, and RNA-binding. It is likely that these modifications influence the activity of p53, which is most effective as a transcription factor when tetrameric; however, the physiological significance of the post-translational modifications remains unknown. The DNA-binding domain is separated from the N-terminal transcriptional activation domain by a region containing a series of proline residues that serve as targets for interactions with modifying proteins (such as signal transduction molecules that contain SH3 binding domains). And the N-terminus serves as a recruiting domain for other transcription factors.

Clues to the role that p53 plays in spermatogenesis may be obtained from observations of cell-specific expression of p53 in the testis. During spermatogenesis, p53 mRNA and protein are present in primary spermatocytes (Almon et al. 1993, Schwartz et al. 1993, and Sjöblom et al. 1996). From these observations, it has been suggested that p53 plays a role in prophase of meiosis.

There is some disagreement about the effects of the p53-deficiency on spermatogenesis. The presence of multinucleated giant cells in p53 knock out testes has been reported (Rotter et al. 1993). These multinucleated cells are thought to result from incomplete meiotic division of pachytene spermatocytes, suggesting that p53 functions in meiotic recombination. Yet, others have indicated that p53 knock out mice on the same genetic background as those used by Rotter et al. (1993) have no evidence of an increased incidence of multinucleated giant cells (Hasegawa et al. 1998).

There are also several reports of alterations in the numbers of certain germ cell populations in p53 knock out mice. Beumer et al. (1998) found that p53 knock out mice have a significant increase in the number of A_1 spermatogonia per Sertoli cell relative to control mice, although there is no increase in spermatogonial stem cells. This change in spermatogonial numbers did not affect the numbers of spermatocytes. In contrast, Yin et al. (1998b) reported lower levels of spontaneous apoptosis in tetraploid germ cells (primary spermatocytes) of p53 knock out male mice and an increase in the relative number of this cell population. And in direct disagreement with both of these studies, Hasegawa et al. (1998) found that there was no change in spermatogonia or spermatocyte numbers and that germ cell apoptosis was no different in p53 knock out mice than in wild type mice.

Some of the differences noted by these various investigators may be due to genetic background, which can drastically influence the expression of some phenotypes. For example, Rotter et al. (1993) reported that p53 knock out mice on a 129 genetic background are infertile, in contrast to C57BL/6 mice which have no evidence of decreased fertility (unpublished observations). Also, differences in the methods used to assess relative numbers of different cell populations could account for variability in these observations.

10.2.2.2 Fas

Fas (CD95/Apo-1) is a cell surface protein in the tumor necrosis factor receptor family and is activated by its ligand, FasL (for a review, see Nagata and Goldstien 1995). Like the other members of this family of apoptosis-inducing proteins, Fas contains cysteine-rich extracellular repeats, apparently necessary for ligand binding, and an intracellular death domain responsible for integrating the extracellular signal to an intracellular transduction event. Crystal structures of the TNF receptor family members suggest that the receptors trimerize upon ligand binding. When activated, the death domain of Fas interacts with caspase 8 which initiates the execution cascade of apoptosis. Mice defective in Fas (*lpr* mice) or FasL (*gld* mice) develop obvious lymphoproliferative and autoimmune disease (Takahashi et al. 1994), and have provided a model for studying the role of Fas in immune system function. As a result, the role of the Fas system in immune cells is particularly well-defined.

The Fas system is also active in testis and participates in the regulation of germ cell numbers during spermatogenesis as well as in the response to injury. Fas ligand is highly expressed in the adult mouse testis, and is localized to Sertoli cells (Lee et al. 1997). Fas is most abundant in primordial germ cells in the fetus, with the highest expression occurring at gestational day 17 and decreasing until birth (Wang et al. 1998a). Fas continues to be expressed in spermatocytes in the adult mouse (Lee et al. 1997). In *gld* mice, which have a point mutation which renders the Fas ligand defective, testis weights and spermatid counts in male mice are significantly increased compared to wild type controls and the apoptotic response to mono-(2-ethylhexyl)phthalate exposure is defective (Richburg et al. 2000). Interestingly, the testis is able to compensate, perhaps by altered transcriptional control, for the Fas gene mutation present in *lpr* mice, a mutation which is known to be leaky (Lee et al. 1997).

10.2.2.3 The bcl-2 Family

The *bcl-2* family of proteins consists of members whose presence promotes cell survival and others whose presence opposes the survival influence of the former (for a review, see Adams and Cory 1998). Those most similar to *bcl-2* in structure inhibit apoptosis by competitively binding adapters needed for caspase activation. Other *bcl-2* family members, such as *bax*, promote apoptosis by displacing anti-apoptotic family members from adapters and freeing the adapter molecules for caspase activation. Pro- and anti-apoptotic family members can hetero- or homodimerize, titrating each other so that the relative cellular concentrations of different family members integrate to influence survival or death. The large family of *bcl-2* proteins are defined by the inclusion of conserved regions termed bcl-2 homology 1 through 4 (BH1–4). These domains control the ability of the proteins to hetero- and homodimerize with other family members. The BH3 domain contains an amphipathic alpha helix and is important in pro-apoptotic family members.

Although many transgenic male mice have spermatogenic defects because of altered gene expression, there are only a few examples in which the gene responsible involves an apoptotic pathway. Transgenic and knock out animal models of the *bcl-2* family members consistently have spermatogenic defects, indicating that the *bcl-2* family is critical in the development of normal spermatogenesis.

The importance of apoptosis to testicular homeostasis is most evident when cell death genes are effectively eliminated, such as in the *bax* knock out mouse (Knudson et al. 1995). Testes of adult male *bax* knock out mice are atrophic with spermatogenic abnormalities ranging from seminiferous tubules devoid of germ cells to tubules with an accumulation of spermatogonia and preleptotene spermatocytes and an absence of more advanced germ cell types (Knudson et al. 1995). The reason for testicular atrophy in an organism lacking a pro-apoptotic gene is not immediately obvious, since one might expect instead a testicular hyperplasia. However, when considered in combination with several studies of transgenic mice overexpressing anti-apoptotic proteins that heterodimerize with bax, a pattern emerges. Mice overexpressing *bcl-2* or *bcl-xl* are also infertile with atrophic testes. Similar to *bax* knock out mice, adult *bcl-2* transgenic male mice lack spermatocytes and spermatids despite an apparently normal appearance and distribution of Leydig cells and Sertoli cells and normal spermatogonial proliferation and apoptosis. Given that the balance between the expression of apoptosis-inhibiting (*bcl-2* and *bcl-xl*) and apoptosis-inducing (*bax*) proteins is critical for cell survival or cell death (Korsmeyer 1995), it is not surprising that transgenic overexpression of *bcl-2* or *bcl-xl* has similar consequences as the functional elimination of *bax* in the testis. But why atrophy instead of hyperplasia? The answer to this important question apparently involves unique features of the first wave of spermatogenesis in which apoptotic spermatogonia express high levels of *bax* protein. How and why inhibition of this early wave of apoptosis affects spermatogenesis throughout the life of the male is not known.

In contrast to the *bcl-2* and *bcl-xl* transgenic mice and *bax* knock out mice in which the balance is tipped toward cell survival, the *bcl-w* knock out mouse lacks a pro-survival factor. These male mice, like their counterparts, are infertile (Print et al. 1998 and Ross et al. 1998), indicating that the loss of an anti-apoptotic factor can be just a detrimental to spermatogenesis as the overexpression of an anti-apoptotic factor (such as *bcl-xl* or *bcl-2*). The complexity of the interaction and importance of these apoptotic genes is evidenced by the anomolous, yet consistent phenotype of atrophic testis in adult transgenic mice. Future studies, examining the expression patterns of these genes during development and how the absence of one factor influences other factors should prove informative.

A further conundrum is the p53 knock out mouse which does not have such dramatic alterations in the development of spermatogenesis, even though both *bax* and p53 are highly expressed in immature testis relative to mature testis (Rodriguez et al. 1997) and p53 is a *bax* gene transcriptional activator (Miyashita et al. 1995). The absence of extensive changes in spermatogenesis in the p53 knock out mouse indicates that there are probably alternate transcription factors for *bax* in the testis.

10.2.2.4 Caspases

A central component of the apoptotic machinery is the system of proteolytic enzymes known as caspases (for a comprehensive review, see Thornberry and Lazebnik 1998). The final steps of any apoptotic signal involve the activation of these proteases and result in the systematic disassembly of the cell. As mammalian homologs of *ced-3*, the caspases are a highly conserved family of cysteine-containing enzymes which target specific amino acid sequences containing aspartate residues. However, the enzymatic activity is not triggered until the protein is cleaved or a co-factor is present. Most caspases contain an aspartate target sequence within their own structure, thereby making each caspase a substrate for other caspases and leading to the generation of a caspase cascade.

Caspases contribute to the characteristic appearance of apoptosis by cleaving target proteins involved in the maintenance of cell structure. For example, the chromatin condensation associated with apoptosis is the result of caspase cleavage of nuclear lamin, which leads to collapse of the nuclear membrane. Additionally, proteins involved in cytoskeletal integrity (gelsolin, focal adhesion kinase, and p21-activated kinase 2) are cleaved by caspases. Caspases also accelerate apoptosis by disabling anti-apoptotic proteins such as *bcl-2* or DNA repair enzymes such as DNA-PK and DNA replication enzymes such as replication factor C.

Intriguingly, although several caspase knock out mouse models exist, no abnormalities in male fertility or spermatogenesis have been reported

(Bergeron et al. 1998, Kuida et al. 1998, Li et al. 1997, Wang et al. 1998b, Woo et al. 1998, Yoshida et al. 1998), suggesting significant redundancy of caspases in the testis or that mice deficient in critical caspases have not yet been engineered.

10.2.3 Experimental Approaches

As an intricately balanced process of proliferation and death, spermatogenesis is susceptible to disruption by a wide variety of insults. For many years, the classic literature in reproductive biology examined the role played by hormones in physiologic control of spermatogenesis. Toxic agents, heat, or radiation often inhibit spermatogenesis by affecting select germ cell populations and, therefore, can be used as probes of specific molecular events. Ultimate targeting of a molecular pathway for experimental investigation involves the examination of animal models in which specific genes have been eliminated or over-expressed. Because of the approaches used in the experimental work to be presented later, this review will concentrate on approaches using animal models and the response to radiation injury in the testis.

10.2.3.1 Knockouts and Transgenics

The up- and down-regulation of apoptotic genes throughout development and during spermatogenesis suggests that apoptosis is regulated in specific germ cells and during certain stages of spermatogenesis. Some of the most convincing evidence for the importance of apoptosis in the maintenance of spermatogenesis comes from knock out and transgenic mice in which important apoptotic genes have been effectively removed or anti-apoptotic genes have been overexpressed. These changes may have little or no apparent effect on spermatogenesis, as in caspase knock out mice, or may lead to severe defects in spermatogenesis, as in transgenic mice over-expressing the *bcl-2* family members.

When an alteration in an apoptotic gene leads to dramatic effects on spermatogenesis, the probable explanation is that the affected gene performs a critical role in testicular apoptosis, either during development or in the maintenance of spermatogenesis. On the other hand, rodent models with abnormal apoptotic genes, but with little or no apparent alteration in baseline spermatogenesis, may manifest dramatic changes in their response to injury. Presumably, although these genes are not necessary for spermatogenesis, they play an essential role in the response to injury. Examples of animal models with near normal baseline spermatogenesis but well documented abnormalities in response to stress include the p53 knock out mouse exposed to radiation (Beumer et al. 1998, Hasegawa et al. 1998, and Hendry et al. 1996) and the *gld* mouse response to phthalate exposure (Richburg et al. 2000).

Potential pitfalls in using transgenic, mutant and knock out mouse models include the unavoidable effects of having an abnormally expressed gene present throughout development and altered in all organ systems. As an example, *gld* mice have a lymphoproliferative autoimmune disease as a result of the loss of the Fas system death pathway in B and T cells (Takahashi et al. 1994). While the effects of this abnormal immune system on testicular function are not known, it has been proposed that FasL expressed in the testis serves to maintain the immune-privilege status of the organ (Bellgrau et al. 1995).

10.2.3.2 Radiation Injury

The effects on spermatogenesis of ionizing irradiation, such as gamma rays, have been known since the early 1900s (Regaud and Blanc 1906) and have been studied extensively (reviewed in Dym and Clermont 1970, Mandl 1964 and Sankaranarayanan 1991). Generally, cell death is the result of damage that results from free radical production during ionizing radiation exposure, DNA being particularly vulnerable to radical-induced alterations. Consequently, the non-dividing supportive Sertoli cells and Leydig cells are relatively unaffected, while selected germ cell populations are rapidly eliminated in response to radiation. The actively dividing spermatogonia are the most susceptible, followed by stem cell spermatogonia, spermatocytes, and the highly resistant spermatids (for a summary of relevant work, see Meistrich et al. 1978). Allan et al. (1987) first described germ cell death resulting from radiation as apoptosis. Hasegawa et al. (1997) later verified, through DNA end-labeling and electron microscopy, that the cell death induced by radiation was indeed due to apoptosis.

From gene and protein expression studies, several genes have been identified as being up-regulated in response to radiation. One such gene is p53, which is normally expressed during meiosis in the preleptotene to early pachytene spermatocytes (Almon et al. 1993, Schwartz et al.

1993, and Sjöblom and Lhädetie 1996). In testis, p53 is closely associated with the nuclear membrane, and upon induction, it is translocated to the nucleus where it may serve as a transcription factor (Yin et al. 1997). Sjöblom and Lhädetie (1996) found that p53 expression is significantly increased in preleptotene and early pachytene spermatocytes within 3 hours of exposure to ionizing radiation, suggesting a role for p53 in meiotic recombination of spermatocytes. In contrast, Beumer et al. (1998) was unable to detect an increase in p53 in pachytene spermatocytes following radiation, observing instead increased expression in preleptotene spermatocytes and spermatogonia. In addition, there is evidence that p53 expression is increased in Sertoli cells in response to radiation (Beumer et al. 1998 and Gobé et al. 1999). Given that the Sertoli cell is intimately involved in signaling germ cell survival and death, it is possible that radiation-induced germ cell death is due, at least in part, to a death signal, such as Fas ligand, produced by the Sertoli in response to increased free radicals within the Sertoli cell.

The functional significance of radiation-induced p53 expression is verified by studies that examine the radiation response in transgenic mice with a targeted disruption of the p53 gene. The absence of p53 in knock out mice leads to a dramatic reduction in the number of apoptotic spermatogonia and spermatocytes in radiation-exposed testes, indicating that not only is p53 expression induced in testicular germ cells in wild type mice by ionizing radiation, but it is necessary for the apoptotic response in germ cells (Hasegawa et al. 1998, Beumer er al. 1998). Interestingly, there are several reports of a lower recovery of spermatogenesis from spermatogonial stem cells in p53 knock out mice (Hendry et al. 1996, Hasegawa et al. 1998, and see our unpublished data below).

10.3 p53/FasL Double Deficient Mouse

A fundamental issue important to any holistic understanding of the role of apoptosis in spermatogenesis is the nature of the interactions between the various apoptotic factors involved. One genetic approach to addressing this issue is to develop animal models with double deficiencies in important apoptotic pathways. By examining the physiology and response to injury in double deficient animals, one can deduce whether the altered pathways are non-interacting or interacting, and if interacting, whether the interaction is additive or synergistic in terms of the effect on spermatogenesis.

We have taken this approach to examine the interaction between p53 and the Fas system in control of germ cell apoptosis. As discussed above, each of these deficient mouse models by itself shows indications that spermatogenesis is altered. Despite conflicting reports, there is evidence that the p53 knock out mouse has an increase in the number of spermatogonia and/or spermatocytes (Beumer et al. 1998, Yin 1998b) and shows a markedly deficient apoptotic response to radiation (Beumer er al. 1998, Hasegawa et al. 1998). The FasL deficient *gld* mouse has a mild testicular hyperplasia and an abnormal apoptotic response to the testicular toxicant, mono-(2-ethylhexyl)phthalate (Lee et al. 1998, Richburg et al. 2000). Although there is evidence that the p53 and Fas systems are closely involved in executing apoptosis, the nature of their interaction has not yet been elucidated. Bennett et al. (1998) demonstrated that p53 induction leads to trafficking of Fas from preformed pools in the Golgi apparatus to the cell surface.

Spermatogenesis, with its high level of physiologic apoptosis and requirement for intimate paracrine interactions, is an ideal model system in which to examine complex interactions among apoptotic pathways. Here, we report our preliminary data on the p53/FasL double deficient mouse, the first *in vivo* model deficient in components of both of these apoptotic systems.

10.3.1 Physiology

Examination of 8 week old wild type (n=5) and p53/FasL double deficient male mice (n=5) indicated that spermatogenesis was normal in the double deficient animal (Table 1). Additionally, male fertility rates of the double deficient mice were similar to those of wild type mice (Table 1), and testis histopathology did not reveal any obvious defects in spermatogenesis (data not shown).

Although baseline testis physiology of the double deficient male mice was similar to that of wild type mice, spleen weights of double deficient mice were greater than those of wild type mice and double deficient mice developed tumors and died within 3–5 months. The increased spleen weights and increased incidence of tumors are attribut-

Table 1. Testis Physiology: Wild Type Mice Compared to Double Deficient Mice

	wild type	double deficient
testis weights	99.55+3.2 mg	98.31+6.2 mg
spermatid head counts	12.3+4.3 million/testis	9.8+3.4 million/testis
tubule diameters	192+20 μm	187+10 μm
fertility rates	77% (n=25)	72% (n=41)

Five male mice from each genotype were killed at 8 weeks of age. Their testes were weighed, and the left testis was processsed for histopathology. Seminiferous tubule cross-sections (100) were measured for each animal. The right testis was homogenized and the number of spermatids were counted on a hemocytometer. Fertility rates are based on mouse colony maintenance records; male mice were mated with 3–5 female mice for two weeks

able to the *gld* and p53 genotypes, respectively, since these observations have been reported for each of the mouse models with single gene deficiencies (Donehower et al. 1992, Reap et al. 1995).

Given the viability of the double deficient males and the comparable baseline physiology of the testis to that of wild type mice, characterization of the response to radiation injury in the double deficient male mice was undertaken.

10.3.2 Response to Radiation Injury

p53 is important in mediating the response to DNA damage induced by irradiation, and it is clear that the Fas system is either directly or indirectly induced by p53. However, how p53 and Fas are related in the response to irradiation is not known. The absence of Fas in the *lpr* mutant mouse is sufficient to significantly inhibit apoptosis in splenic T and B cells derived from gamma irradiated mice, suggesting that, like p53, the Fas system plays an important role in mediating the response to ionizing irradiation (Reap et al. 1997). Meistrich, et al. (1998) and Beumer et al. (1998) found that the normally sensitive differentiating spermatogonia are more resistant to gamma irradiation in the p53 knockout mouse, suggesting that p53 plays an important role in the germ cell apoptotic response to ionizing radiation. Based on these

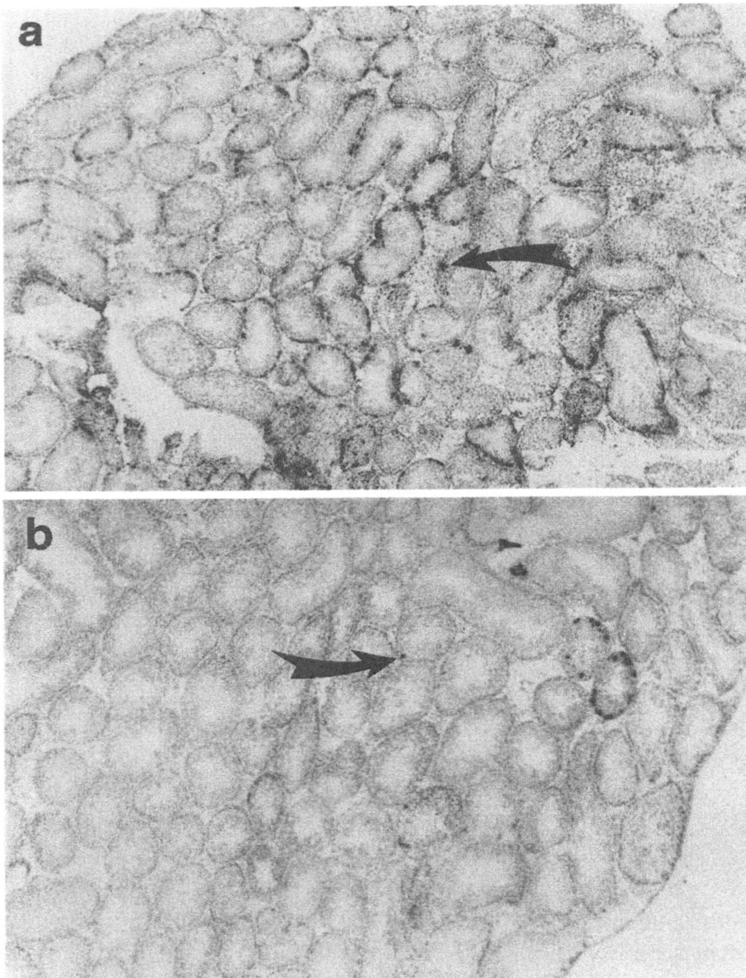

Fig. 1. Comparison of TUNEL-positivity in wild type *versus* double deficient mouse testis. Testicular cross sections were assessed 12 hours following exposure to 5 Gy of x-irradiation to the lower 1/3 of body. **a** Wild type testis showing massive apoptosis. **b** Double deficient testis displaying baseline levels of apoptosis despite radiation exposure. Arrows identify examples of TUNEL-positive cells

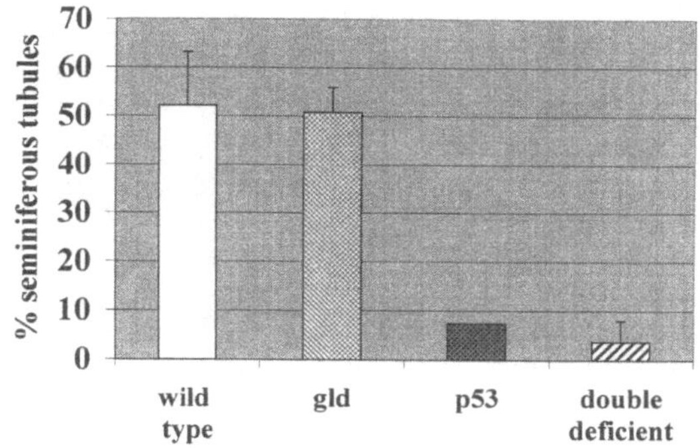

Fig. 2. Quantitation of apoptosis. The number seminiferous tubules having more than 3 apoptotic cells, as determined by TUNEL positivity, were counted 12 hours following exposure to 5 Gy of x-irradiation (see Fig. 1). The number of essentially round seminiferous tubules containing 3 apoptotic cells is expressed as a percentage of the total number of essentially round tubules in a cross-section of the testis. Mice lacking a functional p53 gene show a dramatic decrease in radiation-induced apoptosis relative to mice with p53. Wild type (n=8), *gld* (n=4), p53 knock out (n=1), and double deficient male mice (n=5)

observations, we sought to determine whether male mice deficient in both p53 and FasL would respond to ionizing irradiation any differently than either deficiency alone or the wild type mouse.

Eight week old wild type, *gld*, p53 knock out and double deficient male mice were exposed to 5 Gy of x-irradiation. One control animal of each genotype was sham treated. Twelve hours following treatment, the animals were killed and testes were weighed and collected for histopathology, mRNA expression analysis and determination of apoptosis by the terminal deoxynucleotidyl transferase-mediated deoxy-UTP nick end labeling (TUNEL) technique.

Testis weights were similar in all mice. However, apoptosis, as measured by the number of essentially round seminiferous tubules with greater than 3 TUNEL positive cells, was significantly depressed in double deficient mice compared with wild type or *gld* mice (histopathology, Fig. 1; quantitation, Fig. 2). Semi-quantitative RT-PCR was used to determine that Fas receptor expression was increased in wild type and *gld* mice following the 5 Gy dose, but not in the double deficient or p53 knock out mice.

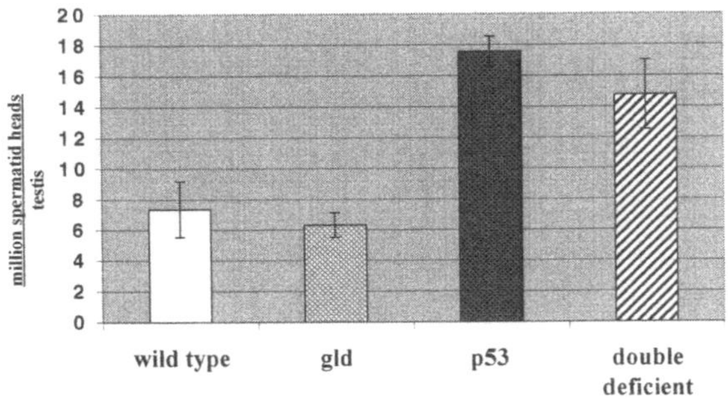

Fig. 3. Quantitation of homogenization-resistant spermatids 29 days post-irradiation. Wild type (n=8), *gld* (n=6), p53 knock out (n=7) and double deficient male mice (n=7) were exposed to 0.5 Gy of x-irradiation to the lower 1/3 of the body, and testes were collected and homogenized 29 days later. Homogenization resistant spermatid heads were counted without knowledge of group identification. Mice lacking a functional p53 gene display a marked increase in spermatids, the progeny of radiation-exposed spermatogonia, relative to mice with p53

In a second experiment, eight week old mice of each genotype were exposed to 0.5 Gy of x-irradiation, a dose that selectively targets the most sensitive population of germ cells, the spermatogonia. Twenty nine days post-dose, animals were killed and their testes were weighed and homogenized. Homogenization resistant spermatid heads, representing the progeny of the cells that had been spermatogonia 29 days before, were counted. Testis weights and spermatid head counts (Fig. 3) were significantly lower in wild type and *gld* mice relative to p53 knock out and double deficient mice. Additionally, spermatid head counts in *gld* mice were statistically lower than wild type mice and spermatid head counts in double deficient mice were statistically lower than p53 knock out mice.

In a third approach to evaluating the effects of the double deficiency on the response to radiation, 8 week old male mice were exposed to 10 Gy of x-irradiation, a dose known to eliminate all germ cells except spermatogonial stem cells (Meistrich et al. 1978). Forty-nine days post-irradiation, mice were killed, and the testes were evaluated histopathologically. The number of seminiferous tubules that had repopulated *versus* the total number of tubule cross-sections was determined as

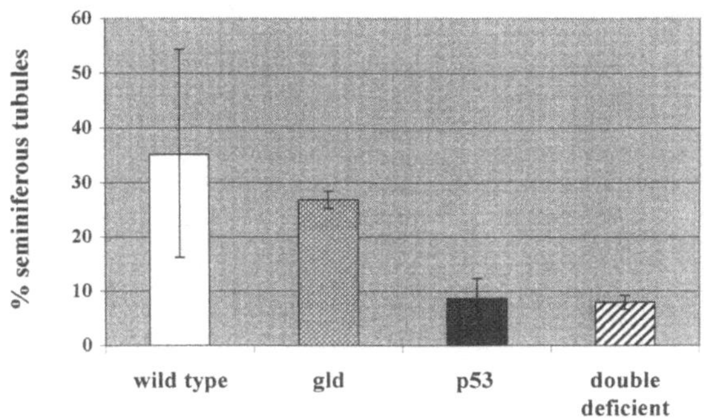

Fig. 4. Repopulation index 49 days post-irradiation. Mice (wild type, n=4; *gld*, n=2; p53 n= 5; double deficient, n=4) were exposed to 10 Gy of x-irradiation to the lower 1/3 of the body. Testes were collected 49 days later and cross sections were examined microscopically for the number of seminiferous tubules containing 3 or more type B spermatogonia as a measure of the ability of stem cells to repopulate the testis. The number of repopulating seminiferous tubules is expressed as a percentage of the total number of tubules assessed. Mice lacking a functional p53 gene have a reduced number of repopulating seminiferous tubules relative to mice with p53

an indication of the ability of the stem cell population to recover after injury (Fig. 4). Both p53 knock out and double deficient mice showed a significant decrease in the number of regenerating tubules relative to *gld* and wild type mice. Although this result is counter-intuitive, based on the results of the exposures to lower doses of radiation and the proposed role of p53 in apoptosis, the results are similar to those previously reported (Hasegawa et al. 1998 and Hendry et al. 1998). One possible explanation for this observation is that stem cells in p53 knock out mice are shorter cycling or have a prolonged radiosensitive phase of the cell cycle, making them more susceptible to the effects of DNA damage. Alternatively, p53 may play a role in cell survival in the stem cell population.

These preliminary experiments have examined p53/FasL double deficient male mice with established spermatogenesis. The double deficient mouse testes are physiologically similar to wild type mouse testes, indicating that neither p53 nor FasL is critical to the baseline apoptosis which occurs during spermatogenesis. The acute apoptotic response to radiation is p53-dependent in male mouse germ cells, and the presence of functional FasL is not necessary for radiation-induced germ cell apoptosis. Radiation exposure induces Fas receptor mRNA expression in the testis, raising the possibility that the Fas system plays a role in the radiation response. The increased Fas receptor mRNA expression in response to ionizing radiation may be, at least in part, p53-dependent.

Overall, the consequences for established spermatogenesis of eliminating these two major apoptotic factors, p53 and FasL, are suprisingly modest. This strongly suggests that germ cell apoptosis is a highly robust process regulated by a multiplicity of redundant pathways.

10.4. Summary and Integration

The high level of physiologic germ cell apoptosis in the testis is apparently used as a mechanism to limit germ cells to a number which can be sustained by Sertoli cells (Huckins 1978). Each Sertoli cell is only able to support, via growth factors and nutrients, a certain number of germ cells. There is also evidence that the Sertoli cell actively limits the number of germ cells by producing death factors, such as FasL, which signal germ cell death (Lee et al. 1997).

Regardless of the mechanisms responsible for cell death, a certain amount of germ cell elimination appears to be necessary for normal spermatogenesis. Two groups have examined the kinetics of germ cell proliferation in separate transgenic mouse models, both over-expressing the anti-apoptotic factor *bcl-2* (Furuchi et al. 1996, Rodriguez et al. 1997). Both groups found inhibition of apoptosis in spermatogonia and resulting testicular atrophy. The *bax* knock out mouse, lacking this important pro-apoptotic factor, is also infertile with atrophic testes (Knudson et al. 1995). On the other hand, both the p53 knock out mouse, lacking the tumor suppresser p53 gene, and the *gld* mutant mouse, lacking a functional FasL, show only a modest testicular hyperplasia, and our preliminary data examining established spermatogenesis in the p53/FasL double deficient mouse indicates that spermatogenesis continues even in the combined absence of both of these factors. This suggests that a certain amount of over-proliferation can be tolerated without deleterious effects on spermatogenesis.

How critical an apoptotic pathway is to spermatogenesis is likely to be determined by at least three important attributes: 1) when the apoptotic factor is expressed during development, 2) in which germ cell types the apoptotic factor predominates, and 3) the extent of disruption of the proliferative/apoptotic balance due to loss of the factor. The markedly adverse consequences for spermatogenesis of deleting *bax* or over-expressing *bcl-2* are probably a result of the importance of this pathway early in development. On the other hand, the p53 and Fas pathways may be used more as fine tuning regulators of established spermatogenesis which are invoked in response to injury.

References

Adams JM, Cory S (1998) The Bcl-2 protein family: arbiters of cell survival. Science 281:1322–1326

Agarwal ML, Taylor WR, Chernov MV, Chernova OB, Stark GR (1998) The p53 network. J Biol Chem 273:1–4

Allan DJ, Gobe GC, Harmon BV (1987) Cell death in spermatogenesis. In: Potter CS (ed) Perspectives on Mammalian Cell Death. Oxford University Press, Oxford, pp 229–258

Allan DJ, Harmon BV, Roberts SA, (1992) Spermatogonial apoptosis has three morphologically recognizable phases and shows no circadian rhythm during normal spermatogenesis in the rat. Cell Prolif 25:241–250

Almon E, Goldfinger N, Kapon A, Schwartz D, Levine AJ, Rotter V (1993) Testicular tissue-specific expression of the p53 tumor supressor gene. Dev Biol 156:107–116

Bartke A (1995) Apoptosis of male germ cells, a generalized or a cell type-specific phenomenon. Endocrinology 136:3–4

Bellamy CO (1997) p53 and apoptosis. Br Med Bull 53:522–538

Bellgrau D, Gold D, Selawry H, Moore J, Franzusoff A, Duke RC (1995) A role for CD95 ligand in preventing graft rejection. Nature 377:630–632

Bennett M, Macdonald K, Chan SW, Luzio JP, Simari R, Weissberg P (1998) Cell surface trafficking of Fas: a rapid mechanism of p53-mediated apoptosis. Science 282: 290–293

Bergeron L, Perez GI, Macdonald G, Shi L, Sun Y, Jurisicova A, Varmuza S, Latham KE, Flaws JA, Salter JC, Hara H, Moskowitz MA, Li E, Greenberg A, Tilly JL, Yuan J (1998) Defects in regulation of apoptosis in caspase-2-deficient mice. Genes Dev 12:1304–1314

Beumer TL, Roepers-Gajadien HL, Gademan IS, van Buul PPW, Gil-Gomez G, Rutgers, DH, and de Rooij, DG (1998) The role of the tumor suppressor p53 in spermatogenesis. Cell Death Differ 5:669–678

Billig H, Furuta I, Rivier C, Tapanainen J, Parvinen M, Hseuh AJ (1995) Apoptosis in testis germ cells: developmental changes in gonadotropin dependence and localization to selective tubule stages. Endocrinology 136:5–12

Blanco-Rodriguez J (1998) A matter of death and life: the significance of germ cell death during spermatogenesis. Int J Androl 21:236–48

Blanco-Rodriguez J, Martinez-Garcia C (1996) Spontaneous germ cell death in the testis of the adult rat takes the form of apoptosis: re-evaluation of cell types that exhibit the ability to die during spermatogenesis. Cell Prolif 29:13–31

Boekelheide K, Lee J, Shipp EB, Richburg JH, Li G (1998) Expression of Fas system-related genes in the testis during development and after toxicant exposure. Toxicol Lett 102–103:503–508

Brinkworth MH, Weinbauer GF, Schlatt S, Nieschlag E (1995) Identification of male germ cells undergoing apoptosis in adult rats. J Reprod Fertil 105:25–33

Coucouvanis EC, Sherwood SW, Carswell-Crumpton C, Spack EG, Jones PP (1993) Evidence that the mechanism of prenatal germ cell death in the mouse is apoptosis. Exp Cell Res 209:238–247

De Rooij DG, Lok, D (1987) Regulation of the density of spermatogonia in the seminiferous epithelium of the Chineses hamster: II. Differentiating spermatogonia. Anat Rec 217: 131–136

Donehower LA, Harvey M, Slagle BL, McArthur MJ, Montgomery CA Jr, Butel JS, Bradley A

(1992) Mice deficient for p53 are developmentally normal but susceptible to spontaneous tumours. Nature 356:215–221

Dym M, Clermont Y (1970) Role of spermatogonia in the repair of the seminiferous epithelium following x-irradiation of the rat testis. Am J Anat 128:265–282

Furuchi T, Masuko K, Nishimune Y, Obinata M, Matsui Y (1996) Inhibition of testicular germ cell apoptosis and differentiation in mice misexpressing bcl-2 in spermatogonia. Development 122:1703–1709

Gobé GC, Harmon B, Leighton J, Allan DJ (1999) Radiation-induced apoptosis and gene expression in neonatal kidney and testis with and without protein synthesis inhibition. Int J Radiat Biol 75:973–983

Hasegawa M, Wilson G, Russell LD, Meistrich ML (1997) Radiation-induced cell death in the mouse testis: relationship to apoptosis. Radiat Res 147:457–467

Hasegawa M, Zhang Y, Niibe H, Terry NH, Meistrich ML (1998) Resistance of differentiating spermatogonia to radiation-induced apoptosis and loss in p53-deficient mice. Radiat Res 149:263–270

Hendry JH, Adeeko A., Potten CS, Morris ID (1996) p53 deficiency produces fewer regenerating spermatogenic tubules after irradiation. Int J Radiat Biol 70:677–682

Huckins C (1978) The morphology and kinetics of spermatogonial degeneration in normal adult rats: an analysis using simplified classification of germinal epithelium. Anat Rec 190:905–926

Knudson CM, Tung KSK, Tourtellotte WG, Brown GAJ, Korsmeyer SJ (1995) Bax-deficient mice with lymphoid hyperplasia and male germ cell death. Science 270:96–98

Korsmeyer SJ (1995) Regulators of cell death. Trends Genet 11:101–105.

Kuida K, Haydar TF, Kuan CY, Gu Y, Taya C, Karasuyama H, Su MS, Rakic P, Flavell RA (1998) Reduced apoptosis and cytochrome c-mediated caspase activation in mice lacking caspase 9. Cell 94:325–337

Lee J, Richburg JH, Shipp EB, Meistrich ML, Boekelheide K (1998) The Fas system, a regulator of germ cell apoptosis, is differentially up-regulated in Sertoli cell *versus* germ cell injury of the testis. Endocrinology 140:852–858

Lee J, Richburg JH, Younkin SC, Boekelheide K (1997) The Fas system is a key regulator of germ cell apoptosis in the testis. Endocrinology 138:2081–2088

Li P, Allen H, Banerjee S, Seshadri T (1997) Characterization of mice deficient in interleukin-1 beta converting enzyme. Cell Biochem 64:27–32

Mandl AM (1964) The radiosensitivity of germ cells. Biol Rev 39:288–371

Meistrich ML, Hunter NR, Suzuki N, Trostle PK, Withers HR (1978) Gradual regeneration of mouse testicular stem cells after exposure to ionizing radiation. Radiat Res 74:349–62

Miyashita T, Reed JC (1995) Tumor suppressor p53 is a direct transcriptional activator of the human bax gene. Cell 80:293–299

Nagata S, Golstein P (1995) The Fas death factor. Science 267:1449–1455.

Oakberg EF, (1956) A description of spermatogenesis in the mouse and its use in analysis of the cycle of seminiferous epithelium and germ cell renewal. Am J Anat 99:391–413

Print CG, Loveland KL, Gibson L, Meehan T, Stylianou A, Wreford N., de Kretser D, Metcalf D, Kontgen F, Adams JM, Cory S (1998) Apoptosis regulator bcl-w is essential for spermatogenesis but appears otherwise redundant. PNAS 95:12424–12431

Reap EA, Leslie D, Abrahams M, Eisenberg RA, Cohen PL (1995) Apoptosis abnormalities of splenic lymphocytes in autoimmune lpr and gld mice. Immunol 154:936–943

Reap EA, Roof K, Maynor K, Borrero M, Booker J, Cohen PL (1997) Radiation and stress- induced apoptosis: a role for Fas/Fas ligand interactions. PNAS 94:5750–5755

Regaud, C. and Blanc, J. (1906) CR Soc Biol 61, 163.

Richburg JH, Nañez A, Williams LR, Embree ME, Boekelheide K (2000) Sensitivity of testicular germ cells to toxicant-induced apoptosis in *gld* mice that express a non-functional form of FasL. Endocrinology, in press

Rodriguez I, Ody C, Araki K, Garcia I, Vassalli P (1997) An early and massive wave of germinal cell apoptosis is required for the development of functional spermatogenesis. EMBO J 16:2262–2270

Ross AJ, Waymire KG, Moss JE, Parlow AF, Skinner MK, Russell LD, MacGregor GR (1998) Testicular degeneration in bcl-w-deficient mice. Nat Genet 18:251–256

Sankaranarayanan K (1991) Ionizing radiation and genetic risks. II. Nature of radiation-induced mutations in experimental mammalian *in vivo* systems. Mutat Res 258:51–73

Schwartz D, Goldfinger N, Rotter V (1993) Expression of p53 protein in spermatogenesis is confined to the tetraploid pachytene primary spermatocytes. Oncogene 8:1487–1494

Selvakumaran M, Lin HK, Miyashita T, Wang HG, Krajewski S, Reed JC, Hoffman B, Liebermann D (1994) Immediate early up-regulation of bax expression by p53 but not TGF beta 1: a paradigm for distinct apoptotic pathways. Oncogene 9:1791–1798

Sjöblom T, Lähdetie J (1996) Expresion of p53 in normal and X-irradiated rat testis suggests a role for p53 in meiotic recombination and repair. Oncogene 12:2499–2505

Takahashi T, Tanaka M, Brannan CI, Jenkins NA, Copeland NG, Suda T, Nagata S (1994) Generalized lymphoproliferative disease in mice, caused by a point mutation in the Fas ligand. Cell 76:969–976

Thornberry NA, Lazebnik Y (1998) Caspases: enemies within. Science 281:1312–1316

Wang R-A, Nakane PK, Koji T (1998a) Autonomous cell death of mouse male germ cells during fetal and postnatal period. Biol Reprod 58:1250–1256

Wang S, Miura M, Jung YK, Zhu H, Li E, Yuan J (1998b) Murine caspase-11, an ICE-interacting protease, is essential for the activation of ICE. Cell 92:501–509

Wang XW, Vermeulen W, Coursen JD, Gibson M, Lupold SE, Forrester K, Xu G, Elmore L, Yeh H, Hoeijmakers JH, Harris CC (1996) The XPB and XPD DNA helicases are components of the p53-mediated apoptosis pathway. Genes Dev 10:1219–1232

Wang XW, Yeh H, Schaeffer L, Roy R, Moncollin V, Egly JM, Wang Z, Freidberg EC, Evans MK, Taffe BG, et al. (1995) p53 modulation of TFIIH-associated nucleotide excision repair activity. Nat Genet 10:188–195
Woo M, Hakem A, Elia AJ, Hakem R, Duncan GS, Patterson BJ, Mak TW (1999) In vivo evidence that caspase-3 is required for Fas-mediated apoptosis of hepatocytes. J Immunol 163:4909–4916
Yin Y, DeWolf WC, Morgentaler A (1997) p53 is associated with the nuclear envelope in mouse testis. Biochem Biophys Res Commun 235:689–694
Yin Y, DeWolf WC, Morgentaler A (1998a) Experimental cryptorchidism induces testicular germ cell apoptosis by p53-dependent and -independent pathways in mice. Biol Reprod 58:492–496
Yin Y, Stahl BC, DeWolf WC, Morgentaler A (1998b) p53-mediated germ cell quality control in spermatogenesis. Dev Biol 204:165–171
Yoshida H, Kong YY, Yoshida R, Elia AJ, Hakem A, Hakem R, Penninger JM, Mak TW (1998) Apaf1 is required for mitochondrial pathways of apoptosis and brain development. Cell 94:739–750

11 Recruitment of p160 Coactivators to Androgen Receptors

M. Parker, C. Bevan

11.1 Does the Androgen Receptor Contain an AF2 Function Identified in Other Nuclear Receptors 166
11.2 The AR Can Interact with Coactivators by Means of LXXLL Motifs 167
11.3 Recruitment of SRC1 to the AR Depends on a Glutamine Rich Region and Not LXXLL Motifs . . 169
11.4 Function of SRC1 As a Coactivator Depends on CBP/p300 Recruitment 170
References 170

Androgens control the proliferation of target cells and a number of physiological responses by means of receptors that function as ligand-dependent transcription factors. Androgen receptors (AR) function either directly by binding to response elements in the vicinity of the promoter or indirectly by modulating the activity of other transcription factors. In common with other nuclear receptors the AR is likely to undergo a characteristic conformational change upon ligand binding that allows the recruitment of cofactors which are required to stimulate transcription of target genes [1, 2, 3]. It first became obvious that nuclear receptors require common cofactors to activate transcription when "squelching" between different receptors was observed: the expression of one active receptor inhibited the activity of a second, implying the existence of an essential limiting cofactor [4].

Numerous potential coactivators have been reported which interact with transcriptionally active receptors in a ligand-dependent manner but the role of many of them is unclear. To date coactivators have been implicated in two distinct steps in target gene activation. Firstly, they play a role in remodelling chromatin, a process that involves the destabilisation of histone-DNA contacts to allow the binding of other transcription factors [5]. Secondly, coactivators are required to recruit the transcription machinery, ultimately RNA polymerase II, to transcribe target genes [6]. This involves the recruitment of a protein complex (Mediator complex) that interacts with one or more subunits of RNA polymerase II. This review will focus on the recruitment of steroid receptor coactivator 1 (SRC1) to the AR, a member of the p160 family of coactivators which appear to play a role in chromatin remodelling.

11.1 Does the Androgen Receptor Contain an AF2 Function Identified in Other Nuclear Receptors

The AR has been shown to stimulate transcription primarily by means of an activation domain near the N-terminus which may be sub-divided into regions [7]. The ligand binding domain is homologous to that in other receptors suggesting a similar structure and function and yet it seems to lack an autonomous ligand-dependent activation domain, AF2, found in other receptors. It is noteworthy that helix 12, which has been shown to be realigned in a number of nuclear receptors and is essential for AF2 activity, is conserved in the AR (Fig. 1). The region comprises an invariant glutamic acid residue and two pairs of hydrophobic amino acids, mutation of which, abolishes the transcriptional activity of the AR [8]. Thus the transcriptional activity of the AR seems to depend not only on the N-terminal AF1 domain but also elements in the LBD. Interestingly, although AF2 activity has not been detected in mammalian cells we found that the LBD fused to a heterologous DNA binding domain was able to stimulate transcription from reporter genes in yeast [8]. Thus, it is conceivable that there is an autonomous activation function in the LBD but its activity may vary in cells depending on the expression of cofactors.

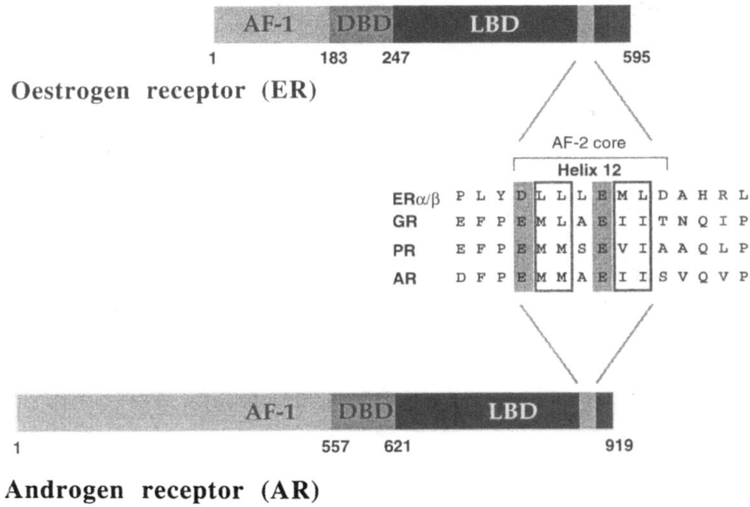

Fig. 1. Organisation of Androgen and Oestrogen Receptors. The domain structure of the receptors is shown together with a sequence alignment of helix in ER and predicted for other steroid hormone receptors. AF-1, activation function 1; DBD, DNA binding domain; LBD, ligand binding domain; AF-2, activation function 2

11.2. The AR Can Interact with Coactivators by Means of LXXLL Motifs

Given the conservation of helix 12 and the potential AF2 activity in the AR we investigated the interaction of the LBD with the putative coactivators, ARA70, SRC1 and TIF2. Using a two-hybrid assay in yeast we found that both SRC1 and TIF 2, and to a lesser extent ARA70, were able to interact in a ligand dependent manner [8]. We then tested their effects on the ability of full-length AR to stimulate transcription from reporter genes in COS-1 cells and found that SRC1 and TIF2, but not ARA70, increased reporter gene transcription. Moreover, the activity of AF2 fused to a heterologous DNA binding domain was stimulated by SRC1 in yeast [8]. Thus we conclude that the SRC1 coactivator has the potential to interact with and potentiate the transcriptional activity of AF2 in a ligand dependent manner.

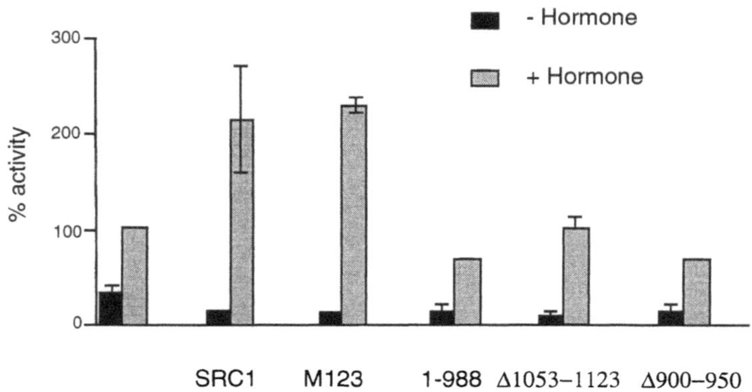

Fig. 2. Ability of SRC1 and mutant versions thereof to stimulate transcription by AR. Transcription from the androgen-responsive reporter pG29GtkCAT was determined in COS 1 cells transfected with androgen receptors in the presence [grey bars] or absence [black bars] of 10nM mibolerone. The effect of SRC1 or a series of mutant versions of SRC1 was then analysed

The recruitment of many coactivators, including SRC1, by nuclear receptors is mediated by hydrophobic interactions between residues located in a cleft formed in their LBDs and helical LXXLL motifs in the coactivators [9, 10, 11, 12]. There are three such motifs in the SRC1family of coactivators whose sequence and spacing is highly conserved and nuclear receptors differ in their preference for different motifs [9]. For example, ER binds preferentially with motif 2 in SRC1 [13]. Mutation of this motif drastically reduces the ability of SRC1 to potentiate transcription by ER and mutation of all three motifs to generate M123 abolished its activity completely. Remarkably, these mutant versions of SRC1 were able to potentiate the transcriptional activity of the AR to a similar extent to that of the wild type protein (Fig. 2). This suggests that, while the LXXLL motifs are capable of interacting with the AR LBD, as shown in yeast two-hybrid assays, an additional site of interaction exists between SRC1 and the AR which is able to compensate for the lack of interaction via LXXLL motifs in M123.

11.3 Recruitment of SRC1 to the AR Depends on a Glutamine Rich Region and Not LXXLL Motifs

To map alternative sites of interaction between SRC1 and the AR we analysed the ability of progressive C-terminal deletion mutants of SRC1 to potentiate the transcriptional activity of the AR in transiently transfected COS-1 cells [8]. The C-terminal residues 1240 to 1399 were found to be dispensable for SRC1 action while deletion to residue 988 resulted in a loss of coactivation activity and a dominant-negative effect, implicating residues 989–1240 in the interaction with the AR. This region is glutamine-rich and shows a high degree of conservation between the p160 proteins. An internal deletion which removes the most highly conserved residues, from 1053 to 1123 was almost inactive indicating that this region may be necessary for interaction with the AR or potentiation of its activity (Fig. 2). Using the yeast two-hybrid system we found that residues 989–1240 were able to interact with the AR in a ligand-independent manner suggesting that the interaction may be mediated by means of AF1. This was confirmed in a second two-hybrid assay performed with the N-terminal AF1 domain and, moreover, deletion of residues 1053 to 1123 from the SRC1 fragment reduced the interaction. Thus we conclude that the major site of interaction between AR and the SRC1 (residues 1052–1123) coactivator is between AF1 and a glutamine rich region in SRC1 which is both necessary and sufficient for the recruitment of this coactivator to the AR.

The N- and C-termini of the AR have been shown to interact upon ligand binding and this interaction is implicated in receptor stabilization, ligand binding and DNA-binding activity [14, 15, 16, 17]. Our data supports the suggestion that SRC1 could promote this interaction by bridging AF1 and AF2 by means of an interaction between the glutamine-rich region and TAU5 (residues 360–494), and between the LXXLL motifs and AF2 [8]. The interaction between SRC1 and AF2 seems to be important, as evidenced by the inactivating effects of helix 12 mutations and yet paradoxically the M123 mutant retains its ability to potentiate transcription. We conclude from our results that the interaction of SRC1 with AF1 by means of the glutamine rich region is the crucial primary step and that additional contacts with AF2 stabilise the interaction.

11.4 Function of SRC1 As a Coactivator Depends on CBP/p300 Recruitment

Finally we have investigated the importance of other regions in SRC1 for its function as a coactivator for AR [8]. Deletion mutants lacking the N-terminal helix-loop-helix, the so-called PAS domain, and a C-terminal activation domain are all capable of potentiating the activity of the AR. However deletion of residues 900–950, which is capable of interacting with CBP/p300, abolishes the activity of SRC1 (Fig. 2). Thus it appears that the ability of SRC1 to potentiate the transcriptional activity of AR depends on the recruitment of the general coactivator CBP/p300. In this way it is postulated that CBP/p300 could facilitate transcription by histone acetylation and may also interact with the basal transcription machinery [18, 19]. It is also interesting to note that CBP/p300 is another candidate for acting as a bridging protein between AF1 and AF2 since it promotes the interaction of these two domains [20, 21].

While SRC1 appears to function as a platform for CBP/p300 it also encodes a number of other functional domains. Thus it encodes histone acetyl transferase activity [22] and it is also capable of recruiting a protein methyl transferase [23]. In addition the HLH and PAS domains are presumably involved in protein dimerisation and/or DNA binding. The challenge for the future is to determine the role of these domains in regulating transcription from specific target genes.

References

1. Mangelsdorf DJ, Thummel C, Beato M, Herrlich P, Schutz G, Umesono K, Blumberg B, Kastner P, Mark M, Chambon P, Evans RE (1995) The nuclear receptor superfamily: the second decade. Cell 83: 835–839
2. Parker MG, White R (1996) Nuclear receptors spring into action. Nature Structural Biology 3: 113–115
3. Glass CK, Rose DW, Rosenfeld MG (1997) Nuclear receptor coactivators. Current Opinion in Cell Biology 9: 222–232
4. Tasset D, Tora L, Fromental C, Scheer E, Chambon P (1990) Distinct classes of transcriptional activating domains function by different mechanisms. Cell 62: 1177–1187
5. Kadanoga JT (1998) Eukaryotic transcription: an interlaced network of transcription factors and chromatin-modifying machines. Cell 92: 307–313

6. Freedman LP (1999) Increasing the complexity of coactivation in nuclear receptor signaling. Cell 97: 5–8
7. Jenster G, van der Korput H, Trapman J, Brinkmann AO (1995) Identification of two transcription activation units in the N-terminal domain of the human androgen receptor. J. Biol. Chem. 270: 7341–46
8. Bevan CL, Hoare S, Claessens F, Heery DM, Parker MG (1999) The AF1 and AF2 domains of the androgen receptor interact with distinct regions of SRC1. Mol Cell Biol 20: 8383–8392
9. Darimont BD, Wagner RL, Apriletti JW, Stallcup MR, Kushner PJ, Baxter JD, Fletterick RJ, Yamamoto KR (1998) Structure and specificity of nuclear receptor-coactivator interactions. Genes Dev 12: 3343–56
10. Mak HY, Hoare S, Henttu PMA, Parker MG (1999) Molecular determinants of the estrogen receptor-coactivator interface. Molecular & Cellular Biology 19: 3895–3903
11. Nolte RT, Wisely GB, Westin S, Cobb JE, Lambert MH, Kurokawa R, Rosenfeld MG, Willson TM, Glass CK, Milburn MV (1998) Ligand binding and co-activator assembly of the peroxisome proliferator- activated receptor-gamma. Nature 395: 137–43
12. Shiau AK, Barstad D, Loria PM, Cheng L, Kushner PJ, Agard DA, Greene GL (1998) The structural basis of estrogen receptor/coactivator recognition and the antagonism of this interaction by tamoxifen. Cell 95: 927–937
13. Kalkhoven E, Valentine JE, Heery DM, Parker MG (1998) Isoforms of steroid receptor coactivator 1 differ in their ability to potentiate transcription by the oestrogen receptor. EMBO J 17: 232–243
14. Doesburg P, Kuil CW, Berrevoets CA, Steketee K, Faber PW, Mulder E, Brinkmann AO, Trapman J (1997) Functional *in vivo* interaction between the amino-terminal, transactivating domain and the ligand binding domain of the androgen receptor. Biochemistry 36: 1052–1064
15. Berrevoets CA, Doesburg P, Sketetee K, Trapman J, Brinkmann AO (1998) Functional interactions of the AF-2 domain core region of the human androgen receptor with the amino-terminal domain and with the transcriptional coactivator TIF-2 (transcriptional intermediary factor-2). Mol Endocrinol 12: 1172–1183
16. Ikonen T, Palvimo JJ, Janne OA (1998) Heterodimerization is mainly responsible for the dominant negative activity of amino-terminally truncated rat androgen receptor forms. FEBS Lett 430: 393–6
17. Langley E, Kemppainen JA, E.M. W (1998) Intermolecular NH2-/carboxy-terminal interactions in androgen receptor dimerization revealed by mutations that cause androgen insensitivity. J Biol Chem 273: 92–101
18. Voegel JJ, Heine MJ, Tini M, Vivat V, Chambon P, Gronemeyer H (1998) The coactivator TIF2 contains three nuclear receptor-binding motifs and

mediates transactivation through CBP binding-dependent and -independent pathways. EMBO J 17: 507–19
19. Torchia J, Rose DW, Inostroza J, Kamei Y, Westin S, Glass CK, Rosenfeld MG (1997) The transcriptional co-activator p/CIP binds CBP and mediates nuclear-receptor function. Nature 387: 677–684
20. Ikonen T, Palvimo JJ, Janne OA (1997) Interaction between the amino- and carboxyl-terminal regions of the rat androgen receptor modulates transcriptional activity and is influenced by nuclear receptor coactivators. J Biol Chem 272: 29821–29828
21. Fronsdal K, Engedal N, Slagsvold T, Saatcioglu F (1998) CREB binding protein is a coactivator for the androgen receptor and mediates cross-talk with AP-1. J Biol Chem 273: 31853–31859
22. Spencer TE, Jenster G, Burcin MM, Allis CD, Zhou J, Mizzen CA, McKenna NJ, Onate SA, Tsai SY, Tsai M-J, O'Malley BW (1997) Steroid receptor coactivator-1 is a histone acetyltransferase. Nature 389: 194–198
23. Chen D, Ma H, Hing H, Koh SS, Huang S-M, Schurter BT, Aswad DW, Stallcup M R (1999) Regulation of transcription by a protein methyltransferase. Science 284: 2174–2177

12 Interaction of Androgens and Oestrogens in Development of the Testis and Male Reproductive Tract

R.M. Sharpe, C. McKinnell, N. Atanassova, K. Williams, K.J. Turner, P.T.K. Saunders, M. Walker, M.R. Millar, J.S. Fisher

12.1	Introduction	173
12.2	Oestrogens and Male Reproductive Development: a Growing Contradiction	176
12.3	Similarities in Effect of Perinatal Administration of Anti-Androgens or Oestrogens	178
12.4	Effect of Neonatal DES Administration to Rats on Induction of Adverse Changes to the Testis and Reproductive Tract in Relation to Androgen Production/Action	181
12.5	Relative Importance of ER-Mediated Effects and Suppression of Androgen Action in Mediating the Adverse Effects of Perinatal DES Treatment on Testicular and Reproductive Tract Development in the Male	187
12.6	What is the Physiological Role of Oestrogens in Male Reproductive Development and Function?	189
12.7	Conclusions	191
References		192

12.1 Introduction

It is accepted that the major driving force that brings about masculinization of the internal reproductive tract and of the external genitalia is testosterone/dihydrotestosterone (androgens). Androgens are also considered to play a central role in masculinization of other organs/tissues in the body, including the brain (Gorski 1996). The main piece of evidence that gave rise to this understanding is the feminization/pseudohermaphroditism that occurs in genotypic males in whom androgen production or action are prevented or are subnormal eg. due to inactivating mutations in the androgen receptor (testicular feminization) (Simpson & Rebar 1995; Hughes & Ahmed 1999). Testicular formation precedes, and is a prerequisite for, masculinization and is controlled via pathways separate to those that control masculinization itself (George & Wilson 1994). Not surprisingly, therefore, gross early development of the testis is not dependent on androgens and genotypic males with 'testicular feminization' develop testes that produce plentiful amounts of testosterone (Simpson & Rebar 1995; Hughes & Ahmed 1999). However, whether development of the testes in such individuals is actually normal is suspect as humans with such disorders are at very high risk of developing testicular germ cell cancer in adulthood and this is thought to stem from abnormal development of the gonocytes in the fetal testis associated with the lack of androgen action (Ottesen et al. 1999). Exactly how the lack of androgen action leads to maldevelopment of the gonocytes is unknown as neither the gonocytes themselves nor the supporting Sertoli cells express androgen receptors (AR) at this stage of development (Majdic et al. 1995).

The discovery three years ago of a second oestrogen receptor (ER), termed ERβ, has re-awakened interest in the possible roles of oestrogens in the male. Comparative studies of the expression of the two ERs, ERα and ERβ, throughout reproductive development in the male has shown that both receptors (but especially ERβ) are expressed widely in the developing testis and reproductive system of the male from fetal life through to adulthood (Fig. 1; see Saunders et al. 1997; Saunders et al.

Fig. 1. Immunoexpression of ERs in the developing reproductive system of the rat. (a) widespread immunoexpression of ERβ in the nuclei of stromal cells surrounding the mesonephric ducts (*) and the Wolffian duct (arrowhead) in late gestation; (b) section of the same block as in (a) but tested with the preimmune serum showing no nuclear staining; (c) widespread immunoexpression of ERα in the nuclei of stromal cells surrounding the mesonephric ducts (*) in late gestation; (d) immunoexpression of ERβ in the nuclei of gonocytes (arrows) and Sertoli cells in the seminiferous cords of the postnatal (day 3) testis of the rat

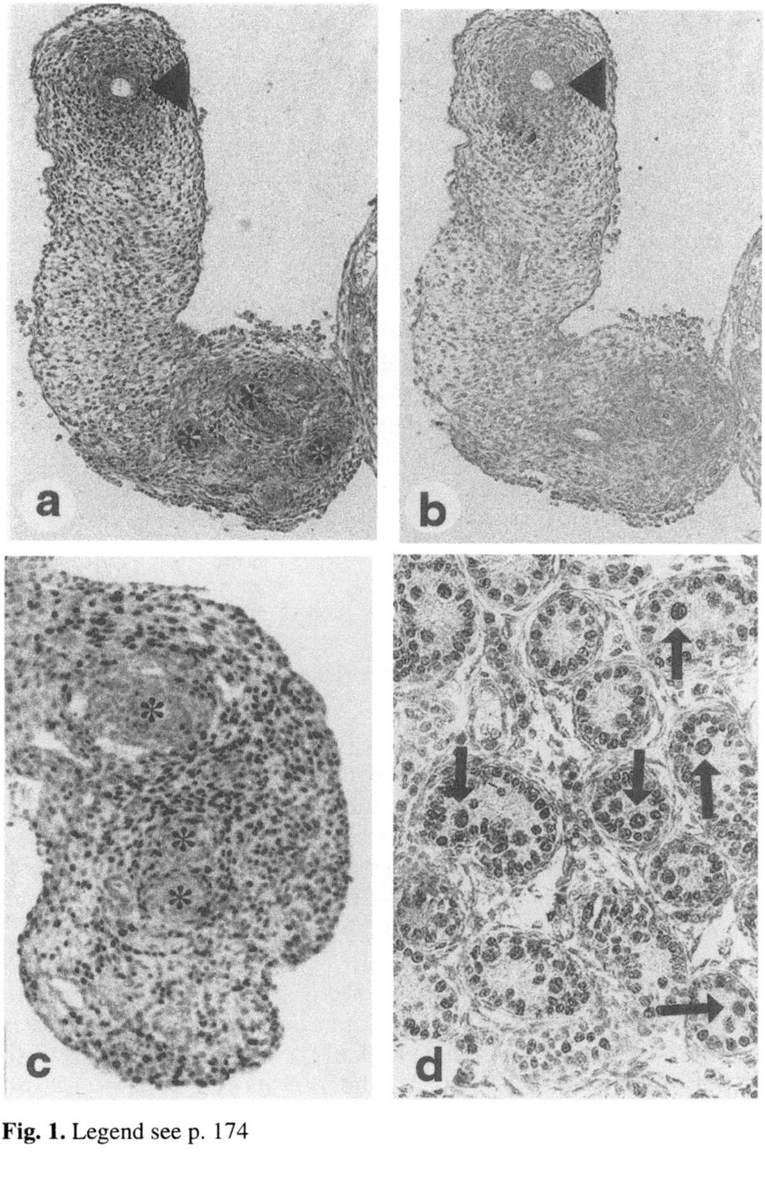

Fig. 1. Legend see p. 174

unpublished data; see also Cooke et al. 1991). Indeed, ERs are expressed more widely than is the AR. Notably, ERβ is expressed in developing Sertoli, Leydig, peritubular and germ cells including particularly heavy expression in gonocytes (Saunders et al. 1997, 1998). With the exception of peritubular cells, none of these cell types express AR in fetal/early postnatal life (Bremner et al. 1994; Majdic et al. 1995). In contrast, outside of the testis in the developing reproductive tract, AR and either or both ERα and ERβ are expressed frequently in the same cell types and it appears to us (though it is unproven) that most, if not all, cells that express AR also express one or both ERs. The pattern of ER expression just outlined appears to be highly conserved in general in species such as the mouse, rat, rabbit, monkey and human (Saunders et al. unpublished data). Such findings have understandably raised the question of the roles that oestrogens may play in development of the testis and reproductive tract, including whether or not some of the effects attributed to androgens might in fact be the result of local conversion of androgens to oestrogens (Sharpe 1998a). The purpose of this paper is to discuss some of our findings and other recently published data that impinge on these questions.

12.2 Oestrogens and Male Reproductive Development: a Growing Contradiction

As just indicated, based on the pattern of expression of ERs it would be predicted that oestrogens must play some role(s) in development of the testis and reproductive tract of the male. The generation of transgenic mice in which either ERα (ERKOs) or ERβ (BERKOs) or aromatase (ArKOs) have been inactivated has given the opportunity to assess the importance of this role(s). Surprisingly, all of these transgenics (homozygotes) develop a grossly normal testis and reproductive tract (Eddy et al. 1996; Krege et al. 1998; CR Fisher et al. 1998; Robertson et al. 1999). The ERKOs and ArKOs do develop disorders of spermatogenesis leading to infertility but both are of relatively late onset (postpubertal in ERKOs; Eddy et al. 1996; late adulthood in ArKOs; Robertson et al. 1999) and in the case of the ERKOs this appears to be an indirect effect resulting from reduced fluid resorption by the efferent ducts (Hess et al. 1997). There is also limited evidence from human males with inactivating mutations in either ERα or aromatase which support the contention that phenotpyic male development is grossly normal (see Sharpe 1998a). These findings suggest strongly that oestrogens are **not** a major player in testicular/reproductive tract development, a conclusion that contrasts starkly with the widespread and conserved distribution of ERs. Further contrast is provided by studies that have investigated the effect of over-exposing the pregnant female or neonatal male to exogenous oestrogens.

Numerous studies, mainly in mice but with supporting human data, have shown that exposure of the pregnant female to high doses of potent oestrogens, such as diethylstilboestrol (DES) or ethinyl oestradiol (EE), will result in an increased incidence of male reproductive disorders, including penile, epididymal and prostatic abnormalities, cryptorchidism, small testes and low sperm counts (Arai et al. 1983; Toppari et al. 1996; Sharpe 1998a, b). Similarly, studies in a range of animals have shown that neonatal administration of the same oestrogens can lead to abnormal development (some of which is permanent) of the rete testis (Aceitero et al. 1998; Fisher et al. 1998, 1999), efferent ducts (Fisher et al. 1998, 1999), epididymis (our unpublished data), prostate (Rajfer & Coffey 1979; Chang et al. 1999) and testis (Sharpe et al. 1998; Atanassova et al. 1999). Until recently, the prevailing view has been that these effects of perinatal oestrogen treatment are indirect effects that are secondary to suppression of gonadotrophin secretion from the pituitary gland (Bellido et al. 1990; Sharpe 1998a). This interpretation was guided by various pieces of supporting evidence but was made without knowledge of the widespread distribution of ERs in the affected target tissues of the male reproductive system. Our own more recent studies, in which we have carefully compared the effects of neonatal oestrogen treatment with those of neonatal gonadotrophin suppression, using a potent GnRH antagonist, suggest that most of the changes induced in oestrogen-treated animals cannot be explained simply by oestrogen-induced suppression of gonadotrophin levels (Sharpe et al. 1998; Atanassova et al. 1999). Most recently it has been shown that elevation of endogenous oestrogen levels, by the generation of transgenic mice that over-express the aromatase gene, results in a broad spectrum of changes to reproductive tract development (Li et al. 1999) which are comparable to those described above in animals administered oestrogens during pregnancy.

Based on the presently available data, there is therefore a major contradiction. On the one hand, the data from transgenic knockout animals suggest a minimal role, if any, for oestrogens in development of the male reproductive system, whereas on the other hand an important role for oestrogens is suggested by studies of the sites of expression of ERs and the widespread changes that are induced in these target tissues by administration of exogenous oestrogens in perinatal life or by the elevation of endogenous oestrogen production. These seemingly opposite conclusions suggest that we are lacking important pieces of information which are necessary for us to be able to reconcile the contradictory data. In this regard, there are various logical possibilities. First, it is possible that the studies of reproductive development in the various transgenic knockout animals have been too superficial and that abnormalities of reproductive development do occur but they are either not grossly evident or they are compensated for, and thus masked, by other changes. This possibility remains to be explored, though it seems certain already that gross phenotypic changes, of the sort that are induced by over-exposure to oestrogens, will not be found. Second, it may be that there is a very close inter-relationship and balance between androgen and oestrogen action during development, with the emphasis always being on keeping the androgen side of the balance in predominance; only when the latter is altered does disruption of male reproductive development occur. It is the latter possibility on which the studies described in this paper will focus.

12.3 Similarities in Effect of Perinatal Administration of Anti-Androgens or Oestrogens

Comparison of the effects in animal studies of administering either an anti-androgen or a potent oestrogen such as DES reveal remarkable similarities in the changes that are induced (Table 1), though it should be noted that the maximum prevalence of most of the disorders in the male offspring would be consistently higher in the animals exposed to the anti-androgen as opposed to the oestrogen. The similarity in phenotypic changes that result from perinatal oestrogen or anti-androgen treatment suggests that common pathways of action may be involved in at least some of these changes. One possibile explanation is offered in Table 2,

Table 1. Comparative effects of altered exposure of the fetus in utero to sex steroids by either the administration of an anti-androgen or by the administration of a potent oestrogen such as DES. Information derived from the following references (Arai et al. 1983; Imperato-McGinley et al. 1992; Toppari et al. 1996; Sharpe 1998b; Mylchreest et al. 1999)

Administration of an anti-androgen	Administration of an oestrogen
Results in increased incidence of:	*Results in increased incidence of:*
At birth:	At birth:
Cryptorchidism	Cryptorchidism
Hypospadias	Hypospadias
Agenesis/disordered development of epididymis and prostate	Epididymal abnormalities
In adulthood:	In adulthood:
Small testes	Small testes
Low sperm counts	Low sperm counts
[Testicular germ cell cancer][a]	Testicular germ cell cancer
	Epididymal cysts
	Prostatic abnormalities

[a] Administration of anti-androgens to animals has not been shown to result in germ cell cancers, but in humans with disorders of androgen production or action (eg. testicular feminisation syndrome) there is an extremely high incidence of such tumours (see Ottesen et al. 1999)

namely that administration of anti-androgens may elevate endogenous oestrogen levels and that this might contribute to some of the adverse reproductive changes in addition to the more straightforward effects resulting from blockade of androgen action. Conversely, oestrogen administration might interfere with androgen production or action (ie. act effectively as an anti-androgen) in addition to activating ER-mediated pathways. These two possibilities are not mutually exclusive and **both** would fundamentally alter the androgen:oestrogen balance by lowering androgen action and elevating oestrogen action (Table 2). Prompted by this possibility, we investigated whether or not neonatal administration of DES at a dose that causes reproductive abnormalities was able to alter androgen production or action. The impetus to undertake this study came from findings of ours in which we showed that neonatal DES

Table 2. Hypothetical postulation that the adverse effects on reproductive tract development in the male after perinatal administration of anti-androgens results from elevation of endogenous oestrogens whilst the effects of perinatal administration of oestrogens result from inhibition of androgen production and/or action

Administration of an anti-androgen	Administration of an oestrogen
Results in a cascade of changes:	*Results in the following changes:*
Blocks central negative feedback effects of androgens, leading to increased of LH secretion	In rats, causes widespread, but transient, reduction in expression of the androgen receptor (AR) in the testis and throughout the developing reproductive tract
Blocks intratesticular negative feedback effects of androgens and this, coupled with increased LH levels, leads to Leydig cell hypertrophy/hyperplasia and increased testosterone production	In rats, retards development of the adult Leydig cell population
Increase in testosterone levels leads to an increase in oestrogen production due to the higher availability of substrate and/or increased induction of the aromatase enzyme	In rats, causes permanent and dose-dependent lowering of circulating testosterone levels
Net result is that androgen action is reduced whilst oestrogen action is increased	Net result is that androgen action is reduced whilst oestrogen action is increased
Comment on potency	Comment on potency
Greater potency of anti-androgens (compared with oestrogens) in causing abnormalities of development of the reproductive system when administered to animals may be because it elevates endogenous oestrogen levels	Less potent than anti-androgens when administered to animals because high doses need to be administered to generate locally high levels in the reproductive system

treatment of rats led to lifelong, dose-dependent suppression of testosterone levels via as yet unknown pathways (Atanassova et al. 1999).

12.4 Effect of Neonatal DES Administration to Rats on Induction of Adverse Changes to the Testis and Reproductive Tract in Relation to Androgen Production/Action

Male rat pups were injected subcutaneously with 10 μg DES in 20μL corn oil on alternate days from day 2–12 inclusive (day of birth = day 1) and compared with littermates that had been injected with either the corn oil alone (controls) or with a potent GnRH anatgonist (GnRHa; antarelix, Europeptides); the latter was administered as two injections, each of 10 mg/kg, on days 2 and 5, a treatment regimen that has been shown to suppress gonadotrophin secretion until ~day 15–20 (Sharpe et al. 1998). The GnRHa treatment group, in which it was assumed that Leydig cell function and development would be impaired because of the lack of LH secretion, allowed us to assess whether this 'anti-androgenic' change resulted in the same effects as those induced by DES.

Injection of DES caused gross structural malformations of the testis/rete testis (Fig. 2), efferent ducts (Figs. 2, 3), epididymis (Fig. 3) and prostate (Fig. 2). The most obvious changes were distension and overgrowth of the rete testis, distension of the efferent ducts with under-development of the apical cytoplasm of the epithelial (resorptive) cells and, in the epididymis and prostate, relative overgrowth of stromal tissue and relative undergrowth of epithelial tissue. None of these changes were observed in animals treated neonatally with GnRHa (Figs. 2, 3) despite the fact that these animals showed similar retardation of testicular growth/weight to the animals treated with DES (Table 3). It is emphasized that all of the tissues/cell types affected adversely by the neonatal DES treatment expressed one or both ERs. In the same tissues, immunoexpression of AR was reduced dramatically in DES-treated rats, and in most cases AR immunoexpression was virtually non-detectable (Figs. 3, 4). In contrast, in GnRHa-treated animals AR immunoexpression was only reduced to a minor extent but the degree of suppression was nowhere near as severe as in the DES-treated group (Figs. 3, 4).

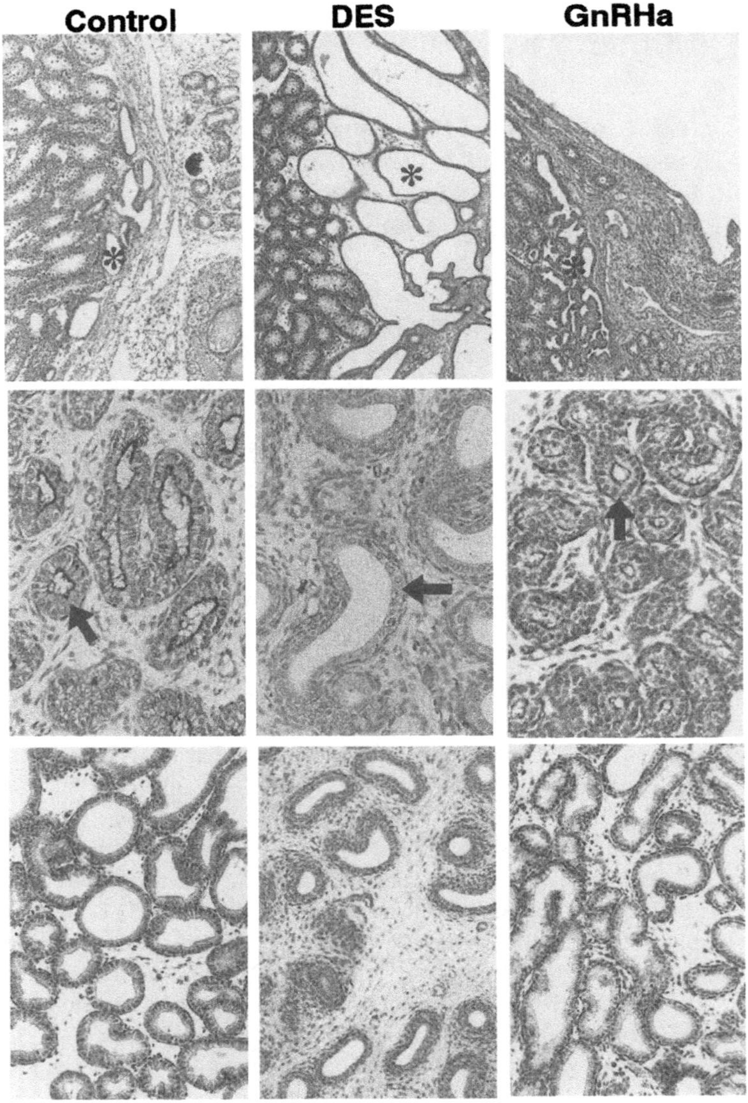

Fig. 2. Legend see p. 183

Table 3. Testis weights and volume of 3β-HSD immunopositive (Leydig) cells per testis at day 18 in rats treated neonatally with vehicle (=controls), 10 μg diethylstilboestrol (DES) or a GnRH antagonist (GnRHa).

Neonatal treatment	Testis weight (mg)	Volume of 3β-HSD immunopositive cells per testis (mm^3)
Control	57.5±11.1	0.10±0.04
DES	26.4±1.9***	0.01±0.01***
GnRHa	26.3±3.2***	0.03±0.01***

Values are the mean±SD for 6 rats per group (3 each from two separate experiments) and cell point-counting (nuceli of 3β-HSD immunopositive cells) followed the stereological procedures reported previously (Atanassova et al. 1999)

***p<0.001, in comparison with control

It has been established previously that androgen withdrawal will lead to loss of detectable immunoexpression of the AR in the testis (Bremner et al. 1994), so to establish whether or not the effect of DES on AR expression was due to suppression of androgen production, immunoexpression of 3β-HSD in the testis (a marker for active Leydig cells) was investigated in control, DES- and GnRHa-treated rats. These showed that both DES and GnRHa treatments led to similar, major reductions in the apparent number of 3β-HSD-positive cells in testis cross-sections (Fig. 4) and quantification of the volume of such cells per testis using standard stereological techniques confirmed this decrease (Table 3).

Fig. 2. Morphology of the rete testis (top row), efferent ducts (middle row) and ventral prostate (bottom row) in rats aged 18 days that had been treated neonatally with either vehicle (=control), 10 μg diethylstilboestrol (DES) or a GnRH antagonist (GnRHa). Note that in DES-treated rats (top centre) the rete testis (asterisks) is grossly distended and overgrown when compared with control (top left) and GnRHa-treated (top right) animals. Note that the lumens of the efferent ducts in DES-treated rats (middle centre) are distended compared with control (middle left) and GnRHa-treated (middle right) animals and that this is associated with a major reduction in height of the epithelial cells (arrows). Note that the ventral prostate in DES-treated rats (bottom centre) exhibits relative overgrowth of stromal compared with epithelial tissue, a change that is not evident in controls (bottom left) or in GnRHa-treated rats (bottom right)

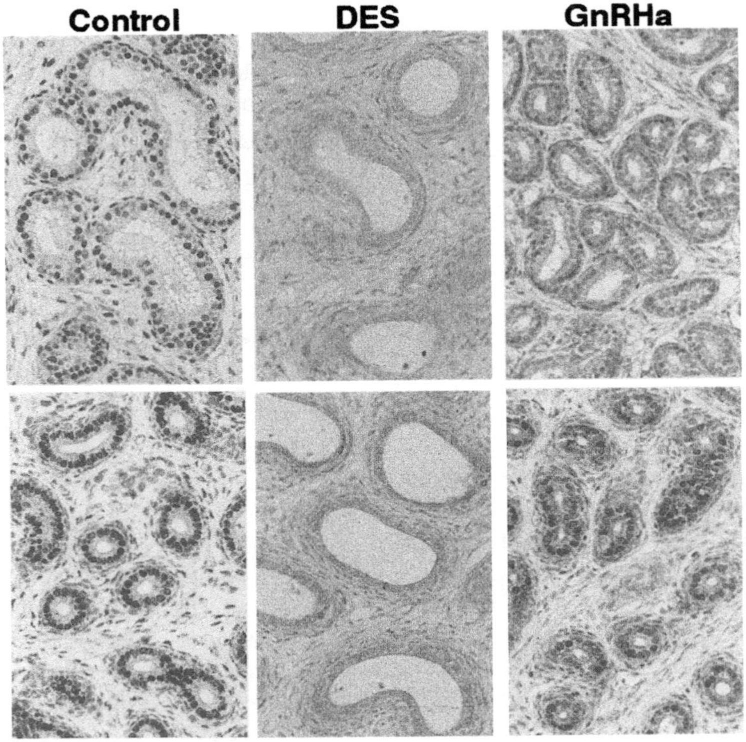

Fig. 3. Tissue morphology and immunoexpression of androgen receptor (AR) in the efferent ducts (top row) and caput epididymis (bottom row) of rats aged 18 days that had been treated neonatally with either vehicle (=control), 10 μg diethylstilboestrol (DES) or a GnRH antagonist (GnRHa). With regard to tissue morphology, note that in DES-treated rats (centre) the lumens of the efferent ducts and epididymal duct are distended compared with control (left) and GnRHa-treated (right) animals and that this is associated with a major reduction in height of the epithelial cells. Note that in controls there is intense nuclear immunoexpression of AR (black) in the epithelial cells of the efferent ducts and epididymal duct as well as in scattered stromal cells (left). This pattern is not altered in GnRHa-treated animals (right) though the intensity of AR immunoexpression appears slightly reduced. In DES-treated animals, AR immunoexpression is virtually non-detectable (middle)

Overall, the impression gained was that testes from animals in the DES- and GnRHa-treated groups contained mainly the residual clumps of fetal Leydig cells and lacked the diffuse, newly differentiating adult-type Leydig cells that predominated in the testes of control animals (Fig. 4). Measurement of testosterone levels in the plasma of control and DES-treated rats at day 18 also showed a significant reduction (Control 0.44±0.14 ng/mL; DES 0.26±0.08; $p<0.001$, means±SD, N=9 and 8 respectively) though the magnitude of the decrease (~40%) was considerably less than the magnitude of the observed decrease in numbers of 3β-HSD-positive cells per testis (~90%); this apparent discrepancy may reflect the established technical problems related to accurate measurement of testosterone levels in fetal/neonatal rats due to the presence of binding proteins and other unidentified 'interfering' factors. In this study, insufficient plasma remained to enable measurement of testosterone levels in the GnRHa-treated rats but the presumption was that a decrease similar to that in DES-treated rats would have been found in view of the similar reduction in Leydig cell numbers (Table 3).

These findings raise numerous questions as to the pathways via which DES treatment causes the reduction in Leydig cell volume, testosterone levels and AR expression. The simplest explanation is that, like the GnRHa treatment, the DES suppresses secretion of LH by the pituitary gland, a possibility that remains to be explored. In this regard, we have shown that FSH secretion is moderately suppressed by the DES treatment though not to levels as low as those found in GnRHa-treated animals (Sharpe et al. 1998; unpublished data). Alternatively, there are bits and pieces of evidence that point to a role of oestrogens in negatively regulating both Leydig cell development and, more acutely, Leydig cell steroidogenesis (see Sharpe 1998a), so it could be that some of the effects observed are a consequence of DES acting via these pathways. It is unclear from the present data whether the dramatic loss of AR immunoexpression in the testis and reproductive tract of DES-treated animals is simply a consequence of reduced androgen availability and action via the AR, or whether other factors might be involved eg. an effect on AR gene transcription or direct interaction of the DES with the AR as proposed by other studies (Yeh et al. 1998). Answers to these questions will require further studies and are probably best addressed by the types of experiments outlined below.

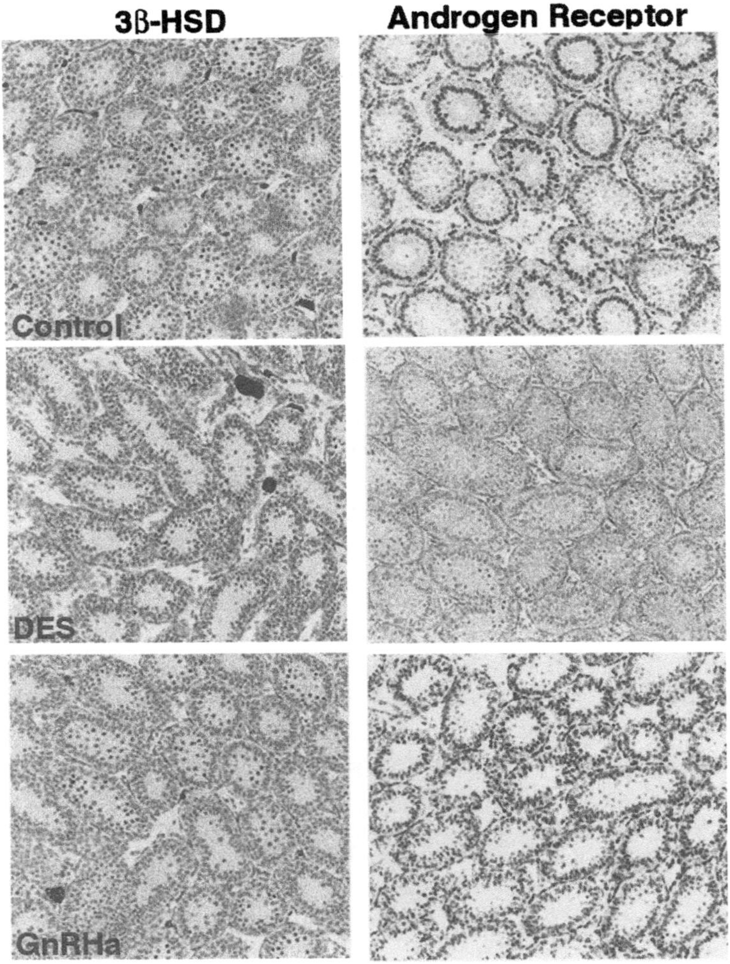

Fig. 4. Legend see p. 187

12.5 Relative Importance of ER-Mediated Effects and Suppression of Androgen Action in Mediating the Adverse Effects of Perinatal DES Treatment on Testicular and Reproductive Tract Development in the Male

In studies that are not reported here, we have shown that administration of lower doses of DES to male rat pups during the same neonatal 'window' fails to induce most of the gross structural changes caused by administering a high dose (10 μg/injection) of DES (Fig. 2) and also fails to cause major loss of AR immunoexpression and suppression of development of Leydig cells. This may be nothing more than coincidence or, on the other hand, it could indicate that suppression of androgen production/action is a key element in determining the severity of many of the effects of DES on the developing male reproductive system. In this regard it is worth emphasizing one key point, namely that it cannot be suppression of androgen action *per se* that mediates the adverse effects of DES treatment as similar suppression of Leydig cell development was induced by GnRHa-treatment without causing any of the gross structural changes that were induced by DES administration during the same neonatal window of time. We have also obtained similar findings in rats treated neonatally with the androgen receptor antagonist, flutamide (unpublished). Therefore, it must be concluded that either the

Fig. 4. Immunoexpression of 3β-hydroxysteroid dehydrogenase (3β-HSD) (left column) and androgen receptor (AR) (right column) in the testes of rats aged 18 days that had been treated neonatally with either vehicle (=control), 10 μg diethylstilboestrol (DES) or a GnRH antagonist (GnRHa). Note that in control rats there are diffuse patches of 3β-HSD-immunopositive (Leydig) cells in the interstitium (top left) whereas in DES-treated (centre left) and GnRHa-treated (bottom left) animals only occasional clumps of 3β-HSD-immunpositive cells are observed which probably represent the residual fetal/neonatal generation of Leydig cells. Note that in controls there is intense nuclear immunoexpression of AR (black) in Sertoli and peritubular myoid cells (top right) and this pattern is not altered in GnRHa-treated animals (bottom right) though the intensity of AR immunoexpression appears slightly reduced in association with delayed development of the seminiferous epithelium. In contrast, in DES-treated animals AR immunoexpression is virtually non-detectable (centre right)

effects of DES treatment are simply the result of its effects via ERs in the target cells or that it is because the high dose of DES upsets the androgen:oestrogen balance by lowering androgen action and elevating oestrogen action simultaneously (see Table 2). Bearing in mind the accepted fact that androgens play a vital role in differentiation and development of the male reproductive tract, it is perhaps most logical to expect that it is the disruption of the androgen:oestrogen balance that is the most likely of these two explanations. If this is the case then it would be predicted that restoration of the androgen:oestrogen balance in DES (10 μg/injection)-treated rats would prevent some (perhaps all) of the gross structural changes induced by the DES treatment alone. Such studies are in progress and the results will be presented at the European Testis Workshop in Saint-Malo.

If the androgen:oestrogen balance is so important for normal development of the male reproductive system it is worthwhile giving further thought to some of the implications that this has, especially in relation to studies of transgenic animals in which oestrogen action is altered. First, the available evidence suggests that the most important factor in the 'balancing act' is the level of androgen action – simply raising oestrogen levels will have relatively little consequence (at least in terms of gross structure) until androgen levels are also reduced. At present, we lack evidence which shows that elevating oestrogen levels whilst maintaining androgen levels will result in gross changes to reproductive tract development, and it could be that, provided androgen levels are maintained at or above normal, oestrogens will be unable to induce major adverse effects. Second, it may be that the lack of gross phenotypic changes to the reproductive systems of male ERKO, BERKO and ArKO mice is a reflection that in these animals all that has been altered is the oestrogen side of the balance, and the direction of the change is downwards rather than upwards as in the studies reported here; moreover, the knockout animals tend to have elevated testosterone levels. It is possible that gross skewing of the androgen:oestrogen balance in favour of androgens is without consequence as far as development of the male reproductive system is concerned. If this thinking is correct then the prediction would be that in knockout mice, in particular the ArKOs, it should still be possible to induce some of the gross structural changes seen with DES in wild-type animals provided that testosterone levels are able to be suppressed to ensure maximum distortion of the normal androgen:oestrogen balance.

12.6 What is the Physiological Role of Oestrogens in Male Reproductive Development and Function?

The presented findings in animals treated with high doses of a potent oestrogen, DES, does not really address the question of the physiological role of oestrogens in the male. As already emphasized, studies of the various transgenic knockout animals suggest that whatever this role might be it is trivial. One clue may be gleaned from our own studies in which we have administered increasingly lower doses of DES to male rat pups during the neonatal period and then studied the testis at the onset of puberty. We have shown that administration on postnatal days 1–12 (injections given on alternate days) of doses of DES ranging from 10 µg down to 0.01 µg/injection causes dose-dependent reduction in Sertoli cell numbers (Atanassova et al. 1999). We have also reported that onset of seminiferous tubule lumen formation and pubertal advancement of spermatogenesis (eg. appearance of spermatocytes) are grossly retarded in animals administered the highest dose (10 µg) of DES (Sharpe et al. 1998). However, study of lower dose DES groups revealed that spermatogenesis is consistently and significantly advanced as determined by the volume of spermatocytes per Sertoli cell (Fig. 5) and also by other markers such as lumen formation and the germ cell apoptotic index (not shown). These findings raise the possibility that oestrogens might play a role in regulating early germ cell development, and in this regard it should be remembered that most spermatocytes and spermatogonia express ERβ (Saunders et al. 1998). This suggestion is supported by a number of reports in the literature that have shown stimulatory effects of oestrogens on gonocyte proliferation in the late fetal testis in the rat (Li et al. 1997), spermatogonial stem cell renewal in the Japananese eel (Miura et al 1999), stimulation of spermatogonial development in the rat (Slowikowska-Hilczer & Kula 1994) and precocious development of spermatogenesis in the human in response to Leydig cell secretion of mainly oestradiol with some testosterone (Kula et al. 1996). Though this data is suggestive and encourages further studies, the lessons learnt from the other studies presented in this paper should also be remembered, namely that androgens and oestrogens probably act in concert. It will therefore be important to assess whether other factors known to be involved in the onset of spermatogenic development in puberty, namely FSH and testosterone, are influenced by the neonatal

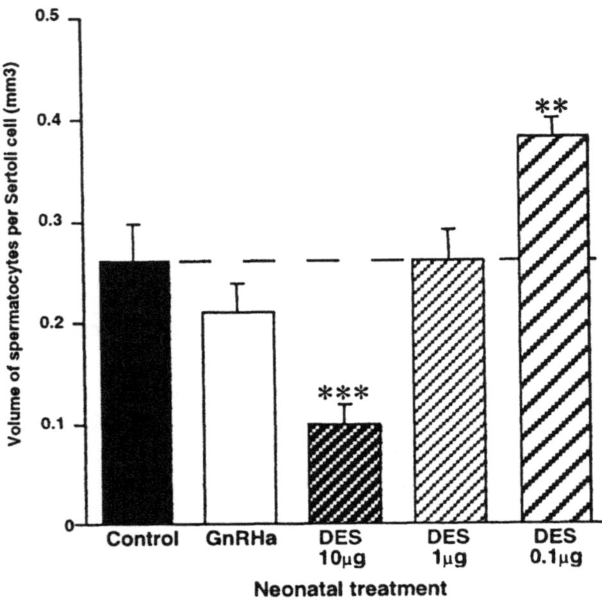

Fig. 5. Effect of neonatal treatment with a GnRH antagonist (GnRHa) or with various doses of diethylstilboestrol (DES) on the onset of spermatogenesis as measured by the volume of spermatocytes per Sertoli cell (means±SEM, N=6) on day 18. The doses of DES shown were administered on alternate days from day 2–12 postnatal and the volumes of spermatocytes and Sertoli cells determined by point-counting based on cell nuclei as reported elsewhere in detail (Atanassova et al. 1999)

administration of low doses of oestrogens as both of these parameters are altered in adulthood in such animals (Atanassova et al. 1999). There are no published studies on the onset of spermatogenesis in ERKO, BERKO and ArKO males, though clearly there is no inordinate delay in puberty in these transgenic lines; however, it will be informative to quantitatively assess spermatogenesis during the normal period of spermatogenic onset in these mice to establish whether or not any changes are detectable that would then point more strongly to a physiological role for oestrogens at this stage.

12.7 Conclusions

The accepted dogma is that androgens regulate development and subsequent function of the male reproductive tract. The present studies do not challenge this view but suggest instead that there may be more to it than this simple one-way process. The emerging data show that elevation of endogenous oestrogen levels (by injecting DES neonatally) is able to induce profound structural changes to the testis and reproductive tract and that for this to happen there must be concomitant suppression of androgen action. Simply suppressing androgen production and/or action on its own will not induce such structural changes, there must also be exposure to oestrogens (though the situation may be different during masculinization of the fetus). In the present studies, high doses of DES were administered to induce structural changes and suppression of androgen action, but the possibility remains that even low doses of oestrogens might also be capable of inducing adverse effects on reproductive tract development if androgen production or action is low or is suppressed for other reasons. Studies to assess this possibility using a treatment combination of anti-androgens and lower doses of DES are in progress. If these generate support for this possibility it would force a serious rethink about our interpretation of many other situations (eg. androgen insensitivity syndromes) which are currently interpreted as resulting solely from the deficiency in androgen action. Irrespective of what proves to be the case, the emerging data suggests that in terms of regulating male reproductive development and function, androgens and oestrogens represent a dual control system that is interlinked in many different ways. Some of these links have been illustrated in the presented studies but there are also others. For example, oestrogen production is dependent on prior androgen synthesis and androgens may also be key regulators of the expression of aromatase and oestrogens might in turn be regulators of 5α-reductase (Sharpe 1998a). The possibility that some oestrogens can interact directly with the androgen receptor (Yeh et al. 1998) is another development that may have to be included in our assessment of how reproductive development is regulated. For certain, the idea that androgens and oestrogens are separate entities with separate, unconnected functions is no longer tenable and future studies must take account of this in their design and interpretation.

Acknowledgements. This work was supported in part by the European Centre for the Ecotoxicology of Chemicals (ECETOC), by AstraZeneca plc and by a Royal Society/NATO fellowship (to NA). We are indebted to Dr R Deghenghi and Europeptides for the generous gift of GnRH antagonist.

References

Aceitero J, Llanero M, Parrado R, Pena E, Lopez-Beltran A (1998) Neonatal exposure of male rats to estradiol benzoate causes rete testis dilatation and backflow impairment of spermatogenesis. Anat Rec 252: 17–33

Arai Y, Mori T, Suzuki Y, Bern HA (1983) Long-term effects of perinatal exposure to sex steroids and diethylstilbestrol on the reproductive system of male mammals. Int Rev Cytol 84: 235–268

Atanassova N, McKinnell C, Walker M, Turner KJ, Fisher JS, Morley M, Millar MR, Groome NP, Sharpe RM (1999) Permanent effects of neonatal oestrogen exposure in rats on reproductive hormone levels, Sertoli cell number and the efficiency of spermatogenesis in adulthood. Endocrinology 140: 5364–5373

Bellido C, Pinilla L, Aguilar R, Gaytan F, Aguilar E (1990) Possible role of changes in postnatal gonadotrophin concentrations in permanent impairment of the reproductive system in neonatally oestrogenized male rats. J Reprod Fertil 90: 369–374

Bremner WJ, Millar MR, Sharpe RM, Saunders PTK (1994) Immunohistochemical localisation of androgen receptors in the rat testis: evidence for stage-dependent expression and regulation by androgens. Endocrinology 135: 1227–1234

Cooke PS, Young P, Hess RA, Cunha GR (1991) Estrogen receptor expression in developing epididymis, efferent ductules, and other male reproductive organs Endocrinology 128:2874–2879

Crisp TM, Clegg ED, Cooper RL, Wood WP, Anderson DG, Baetcke KP, Hoffmann JL, Morrow MS, Rodier DJ, Schaeffer JE, Touart LW, Zeeman MG, Patel YM (1998) Environmental endocrine disruption: an effect assessment and analysis. Environ Health Perspect 106 Suppl 1: 11–56

Eddy EM, Washburn TF, Bunch DO, Goulding EH, Gladen BC, Lubahn DB, Korach KS (1996) Targeted disruption of the estrogen receptor gene in male mice causes alteration of spermatogenesis and infertility. Endocrinology 137: 4796–4805

Fisher CR, Graves KH, Parlow AF, Simpson ER (1998) Characterization of mice deficient in aromatase (ArKO) because of targeted disruption of the *cyp19* gene. Proc Natl Acad Sci USA 95: 6965–6970

Fisher JS, Millar MR, Majdic G, Saunders PTK, Fraser HM, Sharpe RM (1997) Immunolocalisation of oestrogen receptor-α (ERα) within the testis and excurrent ducts of the rat and marmoset monkey from perinatal life to adulthood. J Endocrinol 153:485–495

Fisher JS, Turner KJ, Brown D, Sharpe RM (1999) Effect of neonatal exposure to estrogenic compounds on development of the excurrent ducts of the rat testis through puberty to adulthood. Environ Health Perspect 107: 397–405

Fisher JS, Turner KJ, Fraser HM, Saunders PTK, Brown D, Sharpe RM (1998) Immunoexpression of aquaporin-1 in the efferent ducts of the rat and marmoset monkey during development, its modulation by oestrogens and its possible role in fluid resorption. Endocrinology 139: 3935–3945

George FW, Wilson JD (1994) Sex determination and differentiation. In: The Physiology of Reproduction, 2nd edition. Eds E Knobil, JD Neill, pp3–28. Raven Press, New York

Gorski RA (1996) Androgens and sexual differentiation of the brain. In: Pharmacology, Biology and Clinical Applications of Androgens, Ed S Bhasin et al, New York, Wiley-Liss, pp 159–168

Hess RA, Bunick D, Lee K-H, Bahr J, Taylor JA, Korach KS, Lubahn DB (1997) A role for oestrogens in the male reproductive system. Nature 390: 509–512

Hughes IA, Ahmed SF (1999) Disorders of sexual development and their long-term consequences. In: Fetal programming: influences on development and diseases later in life. Eds PMS O'Brien, T Wheeler, DJP Barker, pp 157–173. Royal College of Obstetricians & Gynaecologists Press, London

Imperato-McGinley J, Sanchez RS, Spencer JR, Yee B, Vaughan ED (1992) Comparison of the effects of the 5α-reductase inhibitor finasteride and the antiandrogen flutamide on prostate and genital differentiation: dose-response studies. Endocrinology 131: 1149–1156

Krege JH, Hodgin JB, Couse JF, Enmark E, Warner W, Mahler JF, Madhabananda S, Korach KS, Gustaffson J-A, Smithies OS (1998) Generation and reproductive phenotypes of mice lacking oestrogen receptor β. Proc Natl Acad Sci USA 95: 15677–15682

Kula K, Slowikowska-Hilczer J, Romer TE, Metera M, Jankowski J (1996) Precocious maturation of the testis associated with excessive secretion of estradiol and testosterone by Leydig cells. Pediatr Pol 71: 269–273

Li H, Papadopoulos V, Vidic B, Dym M, Culty M (1997) Regulation of rat testis gonocyte proliferation by platelet-derived growth factor and estradiol: identification of signalling mechanisms involved. Endocrinology 138; 1289–1298

Li X, Nokkala E, Streng T, Salmi S, Saarinen N, Warri A, Huhtaniemi IP, Makela S, Santti R, Poutanen M (1999) Generation and reproductive phenotypes in transgenic male mice bearing the human ubiquitin C promoter/human P450 aromatase fusion gene. Mol Cell Endocrinol 155: 183 (abstract)

Majdic G, Millar MR, Saunders PTK (1995) Immunolocalisation of androgen receptor to interstitial cells in fetal rat testes and to mesenchymal and epithelial cells of associated ducts. J Endocrinol 147: 285–293

Miura T, Miura C, Ohta T, Nader MR, Todo T, Yamauchi K (1999) Estradiol-17β stimulates the renewal of spermatogonial stem cells in males. Biochem Biophys Res Commun 264: 230–234

Mylchreest E, Sar M, Cattley RC, Foster PMD (1999) Disruption of androgen-regulated male reproductive development by Di (*n*-Butyl) phthalate during late gestation in rats is different from flutamide. Toxicol Appl Pharmacol 156: 81–95

Ottesen AM, Rajpert-de Meyts E, Skakkebaek NE (1999) The role of fetal and infantile development in the pathogenesis of testicular germ cell cancer. In: Fetal programming: influences on development and diseases later in life. Eds PMS O'Brien, T Wheeler, DJP Barker, pp 174–186. Royal College of Obstetricians & Gynaecologists Press, London

Rajfer J, Coffey DS (1979) Effects of neonatal steroids on male sex tissues. Invest Urol 17: 3–8

Robertson KM, O'Donnell L, Jones ME, Meachem SJ, Boon WC, Fisher CR, Graves KH, McLachlan RI, Simpson ER (1999) Impairment of spermatogenesis in mice lacking a functional aromatase (*cyp 19*) gene. Proc Natl Acad Sci USA 96: 7986–7991

Saunders PTK, Maguire SM, Gaughan J, Millar MR (1997) Expression of oestrogen receptor beta (ER beta) in multiple rat tissues J Endocrinol 154:R13–6

Saunders PTK, Fisher JS, Sharpe RM, Millar MR (1998) Expression of oestrogen receptor beta (ERβ) occurs in multiple cell types, including some germ cells, in the rat testis. J Endocrinol 156: R13-R17

Sharpe RM (1998a) The roles of oestrogen in the male. Trends Endocrinol Metab 9: 371–377

Sharpe RM (1998b) Environmental oestrogens and male infertility. Pure Appl Chem 70: 1685–1701

Sharpe RM, Atanassova NN, McKinnell C, Parte P, Turner KJ, Fisher JS, Kerr JB, Groome NB, Mcpherson S, Millar MR, Saunders PTK (1998) Abnormalities in functional development of the Sertoli cells in rats treated neonatally with diethylstilbestrol: a possible role for oestrogens in Sertoli cell development. Biol Reprod 59: 1084–1094

Simpson JL, Rebar RW (1995) Normal and abnormal sexual differentiation and development. In: Principles and practice of endocrinology and metabolism, 2nd edition. Ed. KLBecker, pp. 788–850. Lippincott, Philadelphia

Slowikowska-Hilczer J, Kula K (1994) Comparison between the influence of estradiol benzoate, testosterone propionate and human chorionic gonadotropin on initiation of spermatogenesis in the rat. Ginekol Pol 65: 53–57

Toppari J, Larsen JC, Christiansen P, Giwercman A, Grandjean P, Guillette LJ Jr, Jégou B, Jensen TK, Jouannet P, Keiding N, Leffers H, McLachlan JA, Meyer O, Müller J, Rajpert-De Meyts E, Scheike T, Sharpe RM, Sumpter JS, Skakkebaek NE (1996) Male reproductive health and environmental xenoestrogens. Environ Health Perspect 104, suppl 4: 741–803

Yeh S, Miyamoto H, Shima H, Chang C (1998) from estrogen receptor to androgen receptor: a new pathway for sex hormones in prostate. Proc Natl Acad Sci USA 95: 5527–5532

13 Sertoli Cell Proteins in Testicular Paracriny

D. D. Mruk, C. Yan Cheng

13.1	Introduction	197
13.2	Sertoli Cell Proteins/Factors	199
13.3	Sertoli Cell Paracrine Factors and Germ Cell Function	204
13.4	Sertoli Cell Paracrine Factors and Leydig Cell Function	205
13.5	Potential Roles of Sertoli Cell Paracrine Factors in Germ Cell Movement	205
13.6	Regulation of Junction Assembly – Roles of Sertoli Cell Secreted Paracrine Factors and Other Testicular Factors?	207
13.7	A hypothetical Model Depicting the Participating Molecules and Their Involvement in Germ Cell Movement	209
13.8	Conclusion	209
References		211

13.1 Introduction

Sertoli cells are the irregular-shaped columnar cells in the testis which occupy about 17–19% of the total volume of the seminiferous epithelium in the rat. They extend from the basal to the adluminal compartment of the seminiferous epithelium covering an enormous surface area (for reviews, see Bardin *et al.*, 1988; de Kretser and Kerr, 1988; Jégou, 1993). Due to this unique morphological feature, a single Sertoli cell in the adult rat testis has direct contact with approximately 30–50 germ cells in different stages of development (Russell *et al.*, 1983; Weber *et al.*, 1983; Wong and Russell, 1983) via either anchoring junctions (AJ),

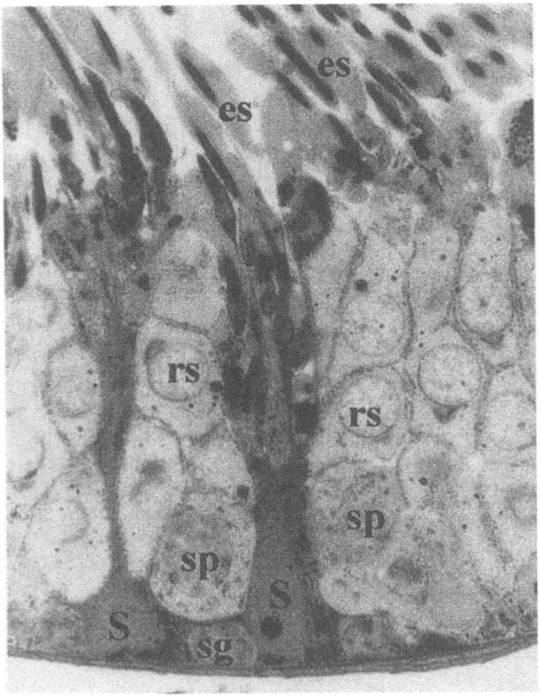

Fig. 1. Cross-section of a seminiferous tubule from an adult rat showing the intimate morphological relationships between Sertoli cells and germ cells at different stages of their development. S, Sertoli cell nucleus; sg, spermatogonium; sp, spermatocyte; rs, round spermatid; es, elongate spermatid. (Courtesy of Dr. Li-ji Zhu)

communicating gap junctions (GJ) or other cellular modifications such as ectoplasmic specializations, an anchoring-like junction, present between elongate spermatids and Sertoli cells. They are also the only somatic cell type in the testis having direct physical contacts and interactions with developing germ cells behind the blood-testis barrier formed by adjacent inter-Sertoli tight junctions (TJ) (except spermatogonia residing outside of the blood-testis barrier in the rat) (for reviews, see de Kretser and Kerr, 1988; Jégou, 1993; Bardin *et al.,* 1988) (Fig. 1). In addition, this barrier also creates a unique microenvironment in which spermatogenesis takes place since virtually all stages of germ cell development, except the mitotic division of spermatogonia and differentiation of type B spermatogonia into preleptotene spermatocytes, are segregated from the systemic circulation. At the basal compartment of the seminiferous epithelium, Sertoli cells also have physical contacts with peritubular myoid cells and have indirect access to cells in the interstitium such as Leydig cells with which interactions occur via paracrine factors. As such, Sertoli cells interact with germ, Leydig, and peritubular myoid cells biochemically via paracrine factors produced by Sertoli cells in addition to the physical contacts between these cells. Moreover, recent studies have demonstrated that paracrine factors released by germ or peritubular myoid cells can also regulate Sertoli cell and testicular functions illustrating the existence of a two-way trafficking between these cells. Since several reviews can be found in the literature dealing with this subject area (see Skinner, 1991; Skinner, 1993; Griswold, 1993; Jégou and Sharpe, 1993), we will limit our discussion in this mini-review on the recent advancements in studying the paracrine regulation of testicular functions via factors produced by Sertoli cells in particular their roles in regulating the turnover of TJ, AJ, and GJ during the process of germ cell movement throughout spermatogenesis.

13.2 Sertoli Cell Proteins/Factors

Sertoli cells are the major secretory cell type in the seminiferous epithelium (for reviews, see Bardin *et al.,* 1988; Griswold, 1993; Jégou, 1993). However, recent studies from this and other laboratories have demonstrated that germ cells are also capable of secreting almost as many proteins as Sertoli cells do when cultured *in vitro* (Chung *et al.,* 1999a). Table 1 summarizes many of the proteins and products synthesized and/or expressed by Sertoli cells in the past two decades many of which have been shown to participate in the paracrine regulation of the testis, such as TGF-β, bFGF, IGF-I, IL-1α, and IL-6. However, the precise physiological roles for many of these proteins in testicular functions are still unknown.

Table 1. Proteins/factors produced and/or expressed by Sertoli cells*

mRNA/Protein	References
Activin-A	Wada *et al.*, 1996; Anderson *et al.*, 1998
Activin-A receptor	Kaipia *et al.*, 1992
cAMP phosphodiesterase genes (PDE): PDE3/IVd and PDE4/IVb	Morena *et al.*, 1995
Adenylyl cyclase	Dym *et al.*, 1991
Androgen binding protein (ABP)	Hagenas *et al.*, 1975; Kierszenbaum *et al.*, 1980; Hall *et al.*, 1990
Androgen receptor (AR)	Blok *et al.*, 1989; 1992; Ree *et al.*, 1999
Anti-Mullerian hormone (AMH)	Vigier *et al.*, 1985; Dutertre *et al.*, 1997
Anti-Mullerian hormone receptor (AMHR)	Dutertre *et al.*, 1997
α-Tubulin	Wrobel *et al.*, 1995; Wolf and Winking, 1996
β_2-Microglobulin	O'Bryan and Cheng, 1997
Basic fibroblast growth factor (bFGF)	Ueno *et al.*, 1987; Mullaney and Skinner, 1992; Lamb, 1993
Basigin	Maekawa *et al.*, 1999
b-myc, c-myc	Lim *et al.*, 1994; Lim and Hwang, 1995
Cadherins: Neural (N-cadherin),) Placental (P-cadherin	Lin and DePhillip, 1996; Chung *et al.*, 1998a
cAMP-dependent protein kinases	Oyen *et al.*, 1988
Cathepsins B, C, D, H, L, and S	Kim and Wright; 1997; Mathur *et al.*, 1997; Chung *et al.*, 1998b
Cellular retinol-binding protein (CRBP)	Ong and Chytil, 1978; Eskild *et al.*, 1991
Ceruloplasmin	Wright *et al.*, 1981; Skinner and Griswold, 1983
c-fos	Hall *et al.*, 1988; Jenab and Morris, 1996; 1997
Claudin-11	Morita *et al.*, 1999
Clusterin	Sylvester *et al.*, 1984; 1991
Collagens: Type I and IV	Skinner *et al.*, 1985; Ulisse *et al.*, 1998;
Complement receptor 1 (CR 1)	Mead *et al.*, 1999
Connexin 33, 43	Tan *et al.*, 1996; Risley *et al.*, 1992
CREB	Waeber *et al.*, 1991; Chaudhary and Skinner, 1999
cyclic AMP	Eikvar *et al.*, 1985
Cyclic dependent kinases (Cdk): Cdc2, Cdk1, Cdk2, Cdk4, Pctaire1, Pctaire-3	Rhee and Wolgemuth, 1995
Cystatin C	Esnard *et al.*, 1992; Tsuruta *et al.*, 1993
Cystic fibrosis transmembrane conductance regulator (CFTR)	Boockfor *et al.*, 1998
Cytochrome P_{450} aromatase (P_{450} arom)	Papadopoulos *et al.*, 1993; Carreau *et al.*, 1999
Decay-accelerating factor (DAF)	Mead *et al.*, 1999
Dynein	Hall *et al.*, 1992; Miller *et al.*, 1999
Endothelin (ET) and enthothelin receptor A (ET-A)	Ergun *et al.*, 1998
Entactin	Ulisse *et al.*, 1998; Mruk and Cheng, unpublished observations

Table 1. Continued

mRNA/Protein	References
Estradiol	Armstrong *et al.*, 1975; Dorrington and Armstrong, 1975; Carreau *et al.*, 1999
Estrogen receptor-β (ERβ)	Saunders *et al.*, 1998; Carreau *et al.*, 1999
F-actin	Pelletier *et al.*, 1995; Wolk and Winking, 1996
Fas ligand (FasL)	Lee *et al.*, 1999; McClure *et al.*, 1999;
Fibroblast growth factor receptors-1 and –4	Le Magueresse-Battistoni *et al.*, 1994; Cancilla and Risbridger, 1998
Follicle stimulating hormone receptor (FSHR)	Means and Huckins, 1974; Orth and Christensen, 1979
Follistatin	Kaipia *et al.*, 1992; Anderson *et al.*, 1998
GATA-1	Ito *et al.*, 1993; Yomogida *et al.*, 1994
Gluthathione S-transferase	Castellon, 1999
G-proteins α_i-1, α_i-2, and α_i-3 subunits	Loganzo Jr and Fletcher, 1992
Haptoglobin	O'Bryan *et al.*, 1997
Hsp27	Welsh *et al.*, 1996
Human Sertoli-cell secreted protein (hSCSP-80)	Papadopoulos, 1991
Inducible cAMP early repressor (ICER)	Walker *et al.*, 1998
Inhibins-α and -β	Klaij *et al.*, 1990; 1992; Pineau *et al.*, 1990; Kaipia *et al.*, 1992
Insulin-like growth factor binding protein (IGFBP)-3	Besset *et al.*, 1996
Insulin-like growth factor-I and -II (IGF-I and -II)	Vannelli *et al.*, 1988; Besset *et al.*, 1996
Integrin subunits α3, α6, β1	Palombi *et al.*, 1992; Salanova *et al.*, 1995; 1998
Intercellular adhesion molecule-1 (I-CAM-1)	De Cesaris *et al.*, 1998
Interferon-γ receptor (INF-γR)	Kanzaki and Morris, 1998
Interleukin-1 receptor type 1	Wang *et al.*, 1998
Interleukin-1α	Stephan *et al.*, 1997; Wang *et al.*, 1998; Jonsson *et al.*, 1999
Interleukin-6 (IL-6)	Stephan *et al.*, 1997; De Cesaris *et al.*, 1998
Kinesin II	Miller *et al.*, 1999
Lactate	Mita *et al.*, 1982
Lactate dehydrogenase A (LDHA)	Santiemma *et al.*, 1987; Boussouar *et al.*, 1999
Laminin	Hadley *et al.*, 1990; Tryggvason, 1993; Ulisse *et al.*, 1998
Leucine-rich primary response gene 1 (LRPR 1)	Slegtenhorst-Eegdeman *et al.*, 1995; 1998
Leukemia inhibitory factor (LIF)	Jenab and Morris, 1997; 1998
Liver regulating protein (LRP)	Gérard *et al.*, 1995
Malate dehydrogenase (MDH)	Santiemma *et al.*, 1987
Membrane cofactor protein (MCP)	Mead *et al.*, 1999

Table 1. Continued

mRNA/Protein	References
Murine transferrin receptor (TfR)	Kissel *et al.*, 1998
Nerve growth factor receptors: low-affinity type and high-affinity type	Persson *et al.*, 1990; Muller *et al.*, 1997
Neurofilament proteins-L (NF-L), NF-M, NF-H	Davidoff *et al.*, 1999
Neuropeptide Y (NPY)	Kanzaki *et al.*, 1996
Nuclear factor (NF)-KB	Delfino and Walker, 1998
Occludin	Saitou *et al.*, 1997; Cyr *et al.*, 1999
Ornithine decarboxylase (ODC)	Madhubala *et al.*, 1987; Weiner and Dias, 1990
Oxytocin	Ivel *et al.*, 1997; Frayne and Nicolson, 1998
p27kip1	Beumer *et al.*, 1999
PEM	Wilkinson *et al.*, 1990
Plasminogen activator inhibitor-1 (PAI-1)	Nargolwalla *et al.*, 1990; Le Magueresse-Battistoni *et al.*, 1998; Bourdon *et al.*, 1998; 1999
Plasminogen activators: tissue-type (t-PA) and urokinase-type (u-PA)	Lacroix and Fritz, 1982
Plectin	Guttman *et al.*, 1999
Prodynorphin and Dynorphin	Collard *et al.*, 1990
Proenkephalin and Enkephalin	Yoshikawa and Aizawa, 1988; Fujisawa *et al.*, 1992
Prolactin receptor (PRL)	Guillaumot and Benahmed, 1999
Prostaglandin D2 synthetase (PGD-S)	Sorrentino *et al.*, 1998; Samy *et al.*, 2000
Proteoglycans: SCI, SCII, and mixed PGs;	Skinner and Fritz, 1985; Phamantu *et al.*, 1995
Putrescine	Tsai and Lin, 1985; Shubhada *et al.*, 1989
Pyruvate	Williams and Foster, 1988;
Reader and Foster, 1990	
Rat and human Sertoli cell-secreted growth factor (r and h SCSGF)	Lamb *et al.*, 1991
REBα	Chaudhary *et al.*, 1999
Retinoblastoma protein (pRb)	Yan *et al.*, 1997
Retinoic acid receptors-α and -β (RAR-α and -β)	Rees *et al.*, 1989; Eskild *et al.*, 1991
Retinoid X receptors (RXR): RXR-α and -β	Mangelsdorf *et al.*, 1992; Γαεμερσ *et al.*, 1998
Retinol dehydrogenase isozymes I and II	Zhai *et al.*, 1997
Retinol-binding protein	Huggenvik and Griswold, 1981; Zhai *et al.*, 1997
Riboflavin carrier protein (RCP)	Bhat *et al.*, 1995; Subramanian and Adiga, 1996
Scinderin	Pelletier *et al.*, 1999
Secreted protein acidic and rich in cysteine	Howe *et al.*, 1988; Cheng, 1990
Seminiferous growth factor	Feig *et al.*, 1980; 1983
Sertoli cell secreted growth factor	Holmes *et al.*, 1986; Shubhada *et al.*, 1996
Sertolin	Mruk and Cheng, 1999

Table 1. Continued

mRNA/Protein	References
Sex-specific antigen (Sxs)	Sanchez *et al.*, 1994
Signal transducers and activators of transcription (STATs): STAT-1 and –3	Jenab and Morris, 1996; 1998
Small heat shock proteins (sHsps)	Welsh and Gaestel, 1998
Somatostatin receptors (SSTs): SST-1, -2, and –3	Zhu *et al.*, 1998
SOX9	Kent *et al.*, 1996; Morais da Silva *et al.*, 1996
Spermidine	Tsai and Lin, 1985; Shubhada *et al.*, 1989
Spermine	Swift and Dias, 1987; Shubhada *et al.*, 1989
Stem cell factor	Jiang *et al.*, 1997; Hakovirta *et al.*, 1999
Steroidogenic Acute Regulatory (StAR) proteins: SCc1, SCc2	Pollack *et al.*, 1997
Sulfated glycoproteins (SGPs): SGP-1 (prosaposin/testibumin) and –2 (clusterin)	Cheng and Bardin, 1986; Igdoura *et al.*, 1996
Superoxide dismutases (SODs)	Mruk *et al.*, 1998
Symplekin	Keon *et al.*, 1996
Testatin	Tohonen *et al.*, 1998
Testin	Cheng *et al.*, 1989; Grima *et al.*, 1998
Testis-determining factor SRY	Salas-Cortes *et al.*, 1999
Testis-specifically expressed cDNAs-1 (tsec-1)	Chen *et al.*, 1997
Thiamin carrier-like protein (TCP)	Subramanian and Adiga, 1999
Tissue inhibitors of metalloproteases-1 and –2	Ailenberg *et al.*, 1991; Grima *et al.*, 1997; Mruk and Cheng, unpublished observations
Tpx-1	Maeda *et al.*, 1998
Transferrin	Skinner and Griswold, 1980; Sigillo *et al.*, 1999
Transferrin receptor	Roberts and Griswold, 1990
Transforming growth factor-α (TGF-α)	Skinner and Moses, 1989; Mullaney and Skinner, 1993
Transforming growth factor-β1 (TGF-β1), -β2, and -β3	Avallet *et al.*, 1994; Le Magueresse-Battistoni *et al.*, 1995
Transforming growth factor-β receptor type II (TGF-βRII)	Suarez-Quian *et al.*, 1989; Olaso *et al.*, 1998
Transient receptor potential (Trp)	Wissenbach *et al.*, 1998
Tryptase	Mruk *et al.*, 1997
Tsx	Cunningham *et al.*, 1998
Vascular cell adhesion molecule-1 (V-CAM-1)	De Cesaris *et al.*, 1998
Vimentin	Wrobel *et al.*, 1995; Rodriquez *et al.*, 1999
Zonula occludens-1 (ZO-1)	Byers *et al.*, 1991; Pelletier *et al.*, 1997
α_2-Macroglobulin	Cheng *et al.*, 1990; Stahler *et al.*, 1991; Braghiroli *et al.*, 1998
β-Catenin	Byers *et al.*, 1994; Chung *et al.*, 1999
β-Synuclein (PNP-14)	Shibayama-Imazu *et al.*, 1998

*This table summarizing many of the proteins and/or factors produced and/or expressed by Sertoli cells is not intended to be exhaustive and readers are strongly encouraged to refer to earlier reviews by Jégou (1993), Bardin et al. (1988), Griswold (1993), Skinner (1993) and the references cited therein, as listed in this table, for additional references. Attempts were made to include more recent articles in the field

13.3 Sertoli Cell Paracrine Factors and Germ Cell Function

Virtually all of the studies performed in the past two decades attempting to delineate the functions of Sertoli cell produced paracrine factors on germ or Leydig cell are the result of *in vitro* studies using primary Sertoli cells co-cultured with germ cells or investigating the effects of Sertoli cell-conditioned media on Leydig cells (for review, see Skinner 1993; Jégou and Sharpe, 1993). While different investigators have selected various end-points to assess the effects of these important testicular paracrine factors, it is not certain if these same effects can be seen *in vivo*. Moreover, there is no reliable *in vitro* assay in which to assess spermatogenesis, as such, it is cautious to note that any apparent effects produced by a particular paracrine factor on the expression of selected genes in either germ and/or Leydig cells co-cultured with Sertoli cells *in vitro* may not correlate with the actual physiological functions *in vivo*.

Utilizing Sertoli-germ cell co-cultures, it was shown that Sertoli cells are needed to (i) maintain germ cell glutathione metabolism (Li *et al.*, 1989), (ii) stimulate germ cell RNA and DNA synthesis (Rivarola *et al.*, 1985), and (iii) induce the appearance of germ cell surface antigens (van der Donk *et al.*, 1986), possibly via paracrine factors secreted by Sertoli cells. Other studies have shown that IGF-I produced by Sertoli cells (Table 1) can bind onto its receptors on meiotic germ cells (Tres *et al.*, 1986; Vannelli *et al.*, 1988) which regulates a number of germ cell functions such as DNA synthesis. Sertoli cells are also known to produce TGF-α and TGF-β (Avallet *et al.*, 1987; Skinner *et al.*, 1989; Skinner and Moses, 1989) possibly involved in the control of germ cell differentiation and meiosis. In addition to these growth factors, Sertoli cells were also shown to express FGF and secrete bFGF in response to FSH (Skinner, 1993). While FGF can stimulate immature Sertoli cell DNA synthesis (Smith *et al.*, 1989; Jaillard *et al.*, 1987), its effect(s) on germ cell function, if any, is not known. In addition, IL-1 produced by Sertoli cells (Gustafsson *et al.*, 1985; Khan *et al.*, 1988; Syed *et al.*, 1988) has been shown to stimulate germ cell DNA synthesis (Pollanen *et al.*, 1989). Other studies have also demonstrated the presence of mRNAs coding for interferons-α and -β in Sertoli, germ, and peritubular myoid cells illustrating the testis is equipped with a sophisticated antiviral defense system (Dejucq *et al.*, 1995; 1998).

13.4 Sertoli Cell Paracrine Factors and Leydig Cell Function

It has long been speculated that Sertoli cells modulate Leydig cell steroidogenesis through a paracrine factor(s) which serves as a local regulatory mechanism to maintain a high androgen level in the testis to facilitate spermatogenesis in addition to participating in the pituitary-testicular axis (Cheng *et al.,* 1988; 1993). As such, numerous attempts have been made to isolate and characterize the Sertoli cell factor(s) modulating Leydig cell steroidogenic functions (Zwain and Cheng, 1994). For instance, it is known that TGF-β inhibits Leydig cell steroidogenesis (Avallet *et al.,* 1987). In addition, a recent report has shown that the factor isolated from Sertoli cell-conditioned medium having potent stimulatory effects on Leydig cell steroidogenesis is possibly a cysteine protease-metalloprotease inhibitor complex of cathepsin L-TIMP-1 (Boujrad *et al.,* 1995). Studies have also illustrated that TGF-β inhibits DNA synthesis in immature rat Leydig cells *in vitro* (Khan *et al.,* 1999) suggesting TGF-β produced by Sertoli cells may affect other Leydig cell functions, in addition to steroidogenesis. Moreover, Sertoli cells are known to produce a LHRH-like factor that affects Leydig cell steroidogenesis since receptors for LHRH are found on Leydig cells (Sharpe *et al.,* 1981; Sharpe and Fraser, 1980). These results seemingly suggest that the local regulatory mechanisms of paracrine factors modulating Leydig cell steroidogenesis are accomplished by an array of molecules some of which have opposite biological effects: stimulatory and inhibitory. It is possible that through this unique arrangement, a high but uninterrupted level of androgens can be maintained within the seminiferous epithelium essential for germ cell development and maturation.

13.5 Potential Roles of Sertoli Cell Paracrine Factors in Germ Cell Movement

Needless to say, the most significant morphological event during spermatogenesis is meiosis in which a single diploid cell (type B spermatogonium) must divide and differentiate into 4 haploid cells (spermatids). However, the timely movement of developing germ cells across the seminiferous epithelium is equally important to the completion of spermatogenesis. Without a precise mechanism regulating the timely translocation of germ cells from one site to another during their development, mature spermatids (spermatozoa) cannot be released into the tubular lumen during spermiation which would result in male infertility. Moreover, while this process is important, virtually no studies have been conducted to tackle this biological event.

Earlier morphological analyses have suggested that the movement of germ cells across the seminiferous epithelium is largely dependent upon Sertoli cells since germ cells lack the necessary cytoskeletal elements and locomotive components for movement, yet the regulatory mechanism employed by Sertoli cells triggering the cascade of events leading to germ cell movement is also not entirely known. During the past several years, studies from our laboratory have shown that the cascade of events leading to germ cell movement is likely composed of intermittent phases of junction disassembly and reassembly. Without this timely disruption of inter-Sertoli and Sertoli-germ cell junctions, which in turn allows for the protrusion of Sertoli cell cytoplasmic processes to assist in the translocation of germ cells from one site to another, no germ cell movement can take place. In addition, the disassembled junctions must be reassembled in a highly coordinated manner so that germ cells can remain attached onto the seminiferous epithelium and the integrity of the blood-testis barrier can be maintained. While it is virtually impossible to study the events of junction disassembly and assembly even with the use of staged tubules since multiple biochemical events are taking place concurrently many of which are unrelated to cell movement, it is possible to study the events of junction reassembly using primary cultures of Sertoli cells or co-cultures of Sertoli-germ cells since it has been shown that different types of junctions including TJ, AJ, and GJ are known to form between testicular cells when cultured *in vitro*.

When Sertoli-germ cells were co-cultured *in vitro*, AJ such as desmosome-like junctions are formed within 24–48 hr (Cameron and Muffly, 1991; Enders and Millette, 1988). Using such a simple *in vitro* model, we have demonstrated that the addition of germ cells into primary Sertoli cells cultured in bicameral units induces a transient but significant increase in serine protease production in the apical chamber at 3 hr at the time of germ cell attachment onto Sertoli cells but preceding the establishment of Sertoli-germ cell junctions seemingly suggesting that cell surfaces need to be prepared before junctions can be formed

(Mruk *et al.*, 1997). Studies using semi-quantitative RT-PCR were able to confirm the transient but significant expression of uPA (urokinase type plasminogen activator, a serine protease), tryptase (also a serine protease), in addition to cathepsin L (a cysteine protease). More important, recent studies using fluorescein-labeled germ cells co-cultured with Sertoli cells in the presence of aprotinin (a serine protease inhibitor) and α_2-macroglobulin (a Sertoli cell secretory product and non-specific protease inhibitor, see Table 1) have illustrated that the binding of germ cells onto Sertoli cells can be cnhanced in the presence of these protease inhibitors by 24 hr, but not earlier when germ cells are binding onto Sertoli cells, clearly suggesting the events of junction assembly (germ cell binding is a prerequisite of the subsequent junction assembly event) requires the participation of non-junction-associated molecules. Interestingly, recent studies have shown that the timely expression of cathepsin L can be regulated by bFGF illustrating this particular paracrine factor may play a key role in modulating these events (Mruk and Cheng, unpublished observations).

Other studies utilizing Sertoli cells cultured *in vitro* have also demonstrated that transient expressions of ZO-1 (zonula occludens-1, a TJ-associated protein), N-Cad (N-cadherin, an AJ-associated protein), and Cx33 (connexin33, a GJ-associated protein) are associated with the assembly of TJ, AJ, and GJ, respectively, *in vitro* (Chung *et al.*, 1999b; Wong *et al.*, 2000) further illustrating that the events of germ cell movement, which require intermittent phases of junction disassembly and reassembly, are regulated by an array of molecules. Work is now in progress to determine if the timely expression of these junction-associated molecules is regulated by Sertoli cell secreted paracrine factors, such as TGF-α and -β.

13.6 Regulation of Junction Assembly – Roles of Sertoli Cell Secreted Paracrine Factors and Other Testicular Factors?

The identity of the regulator(s) that induces these transient and significant changes in target genes' expression when inter-Sertoli or Sertoli-germ cell junctions are being assembled *in vitro* as described above is not known. It is possible that cytokines play a role since cytokines such as bFGF and TGF-β have been shown to affect Sertoli cell PA expression and/or production *in vitro* (Nargolwalla *et al.,* 1990; Jaillard *et al.,* 1987). TGF-β, bFGF, and EGF can also affect Sertoli cell PAI-1 (plasminogen activator inhibitor-1, a serine protease inhibitor) expression *in vitro* (Le Magueresse-Battistoni *et al.,* 1998). In addition, numerous studies have demonstrated that changes in cell shape and reorganization of the cytoskeleton pertinent to cell/substratum adhesion and subsequent cell motility is regulated by EGF through EGF receptor (EGFR)-mediated signaling pathways (for review, see Wells *et al.,* 1999), implicating the direct participation of cytokines in cell movement. Other studies have shown that both Sertoli and germ cells express and/or produce a variety of cytokines such as TGF-β and-α, NGF, FGF, INF-α, -β, and - γ, and TNF-α which play multiple roles in regulating testicular cell functions (for reviews, see Griswold, 1995; Jégou, 1993; Lamb, 1993). Moreover, many of the molecules associated with testicular cell junctions such as occludin, ZO-1, and the N-Cad/β-catenin complex are known to take part in junction assembly *in vitro* via their phosphorylation and dephosphorylation by specific protein kinases and tyrosine phosphatases, respectively. The extent of phosphorylation of these junction-associated molecules can also regulate the status and permeability of cell junctions. Recent studies in the blood-brain barrier have shown that the state of phosphorylation of ZO-1, N-Cad, β-catenin, and p120 are regulated by the intracellular signaling processes participating in the regulation of TJ permeability (for review, see Rubin and Staddon, 1999). For instance, both ZO-1 and β-catenin are targets for tyrosine kinases and can be phosphorylated at the tyrosine residue, in turn increasing TJ permeability (Staddon *et al.,* 1995). p120, an E-Cad/β-catenin complex-associated protein which is a Src substrate, and p100, a p120-related protein (Reynolds *et al.,* 1994; Shibamoto *et al.,* 1995), are both known to become phosphorylated on tyrosine, serine, and threonine residues and these phosphorylated proteins then become sensitive to a variety of stimuli that affect TJ permeability (Ratcliffe *et al.,* 1997). Another study using multiple inhibitors has demonstrated that G-proteins, phospholipase C, adenylate cyclase, protein kinase C, and calmodulin participate in the regulation of the TJ assembly and sealing in MDCK epithelial cells *in vitro* (Balda *et al.,* 1991) seemingly suggesting that the regulation of these junctions in the testis is likely to require a coordinated mechanism involving numerous molecules.

13.7 A Hypothetical Model Depicting the Participating Molecules and Their Involvement in Germ Cell Movement

Figure 2 illustrates a hypothetical model of germ cell movement within the testis and the potential role of paracrine factors regulating this cellular phenomenon. Germ cell movement is likely composed of continuous but intermittent phases of junction disassembly, followed by the protrusion of Sertoli cell cytoplasmic processes that assist in germ cell translocation from one site to another (Phase I), and junction reassembly (Phase II). Junction disassembly is likely to be initiated by a cytokine(s) released from either Sertoli and/or germ cells which in turn triggers the production of proteases (pathway 1). Proteases cause junction cleavage which permits germ cell movement with the help of the Sertoli cells which generate protruding cytoplasmic processes, possibly functioning as depicted in Fig. 2 (pathway 2). Thereafter, a cytokine(s) released once again from either Sertoli and/or germ cells induces the production of protease inhibitors (pathway 3). Protease inhibitors limit the further action of proteases, thereby protecting the integrity of the testis and the blood-testis barrier. This is followed by the production of new junction-associated molecules under the influence of cytokines (pathway 4) so that the inter-Sertoli and Sertoli-germ cell junctions can be reassembled. Although this biochemical model of germ cell movement will continue to be revised and updated, it has now become the basis of our current and future investigations.

13.8 Conclusion

This brief review summarizes some of the recent developments in the field in particular the roles of paracrine factors in regulating testicular function with emphasis on the possible involvement of paracrine factors in the event pertinent to the movement of germ cells within the seminiferous epithelium. It is anticipated that the next decade will bring about new and exciting developments on the mechanism of germ cell migration within the seminiferous epithelium, along with the participating molecules.

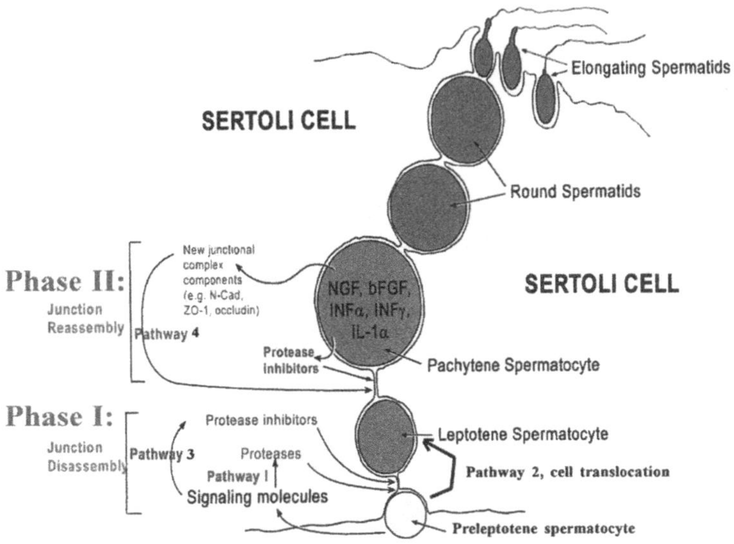

Fig. 2. A hypothetical model depicting the two phases of germ cell movement, which consist of intermittent phases of junction disassembly/cell protrusion (Phase I) and reassembly (Phase II). Phase I consists of the induction of cytokines and production of proteases (pathway 1) needed to cleave inter-Sertoli and Sertoli-germ cell junctions to allow for germ cell movement. After germ cells complete their translocation with the help of Sertoli cell cytoplasmic processes that generate the protruding force (pathway 2), cytokines will be released to activate the production of protease inhibitors (pathway 3) to limit the further action of proteases. Phase II involves the production of new junctional complex components (pathway 4) under the influence of cytokines and/or protein tyrosine kinases/protein tyrosine phosphatases to re-assemble the disrupted junctions. This is a highly schematic model since many of the participating molecules in these events remain to be identified and studied.

References

Ailenberg M, Stetler-Stevenson WG, Fritz IB (1991) Secretion of latent type IV procollagenase and active type IV collagenase by testicular cells in culture. Biochem J 279:75–80

Anderson RA, Evans LW, Irvine DS, McIntyre MA, Groome NP, Riley SC (1998) Follistatin and activin A production by the male reproductive tract. Human Reprod 13:3319–3325

Armstrong DT, Moon YS, Fritz IB, Dorrington JH (1975) Synthesis of estradiol-17β by Sertoli cells in culture: stimulation by FSH and dibutryl cyclic AMP. Curr Top Mol Endocrinol 2:85–96

Avallet O, Vigier M, Leduque P, Dubois PM, Saez JM (1994) Expression and regulation of transforming growth factor-β1 messenger ribonucleic acid and protein in cultured porcine Leydig and Sertoli cells. Endocrinology 134:2079–2087

Avallet D, Vigier M, Perrard-Sapori MH, Saez JM (1987) Transforming growth factor beta inhibits Leydig cell functions. Biochem Biophys Res Commun 146:575–581

Balda MS, Gonzalez-Mariscal L, Contreras RG, Macias-Silva M, Torres-Marquez ME, Garcia Sainz JA, Cereijido M (1991) Assembly and sealing of tight junctions: possible participation of G-proteins, phospholipase C, protein kinase C and calmodulin. J Membrane Biol 122:193–202

Bardin CW, Cheng CY, Musto NA, Gunsalus GL (1988) The Sertoli cell. In: The Physiology of Reproduction. Vol 1. Eds. Knobil E, Neill JD, Ewing LL, Greenwald GS, Markert CL and Pfaff DW. New York, Raven Press, pp. 933–974

Besset V, Le Magueresse-Battistoni B, Collette J, Benahmed M (1996) Tumor necrosis factor-α stimulates insulin-like growth factor binding protein 3 expression in cultured porcine Sertoli cells. Endocrinology 137:296–303

Beumer TL, Kiyokawa H, Roepers-Gajadien HL, Van der Bos LA, Lock TM, Gademan IS, Rutgers DH, Koof A, De Rooij DG (1999) Regulatory role of p27kip1 in the mouse and human testis. Endocrinology 140:1834–1840

Bhat KG, Malhotra P, Karande AA, Adiga PR (1995) Immunohistochemical localization of riboflavin carrier protein in testicular cells of mammals. Indian J Exp Biol 33:12–16

Blok LJ, Hoogerbrugge JW, Themmen AP, Baarends WM, Post M, Grootegoed JA (1992) Transient down-regulation of androgen receptor messenger ribonucleic acid (mRNA) expression in Sertoli cells by follicle stimulating hormone is followed by up-regulation of androgen receptor mRNA and protein. Endocrinology 131:1343–1349

Blok LJ, Mackenbach P, Trapman J, Themmen AP, Brinkmann AO, Grootegoed JA (1989) Follicle stimulating hormone regulates androgen receptor mRNA in Sertoli cells. Mol Cell Endocrinol 63:267–271

Boockfor FR, Morris RA, De Simone DC, Hunt DM, Walsh KB (1998) Sertoli cell expression of the cystic fibrosis transmembrane conductance regulator. Am J Physiol 274:C922–930

Boujrad N, Ogwuegbu SO, Garnier M, Lee CH, Martin BM, Papadopoulos V (1995) Identification of a stimulator of steroid hormone synthesis isolated from testis. Science 268:1609–1612

Bourdon V, Defamie N, Fenichel P, Pointis G (1999) Regulation of tissue-type plasminogen activator and its inhibitor (PAI-1) by lipopolysaccharide-induced phagocytosis in a Sertoli cell line. Exp Cell Res 247:367–372

Bourdon V, Lablack A, Abbe P, Segretain D, Pointis G (1998) Characterization of a clonal Sertoli cell line using adult PyLT transgenic mice. Biol Reprod 58:591–599

Boussouar F, Grataroli R, Ji J, Benahmed M (1999) Tumor necrosis factor-γ stimulates lactate dehydrogenase A expression in porcine cultured Sertoli cells: mechanisms of action. Endocrinology 140:3054–3062

Braghiroli L, Silvestrini B, Sorrentino C, Grima J, Mruk D, Cheng CY (1998) Regulation of α_2-macroglobulin expression in rat Sertoli cells and hepatocytes by germ cells in vitro. Biol Reprod 59:111–123

Byers S, Graham R, Dai HN, Hoxter B (1991) Development of Sertoli cell junctional specializations and the distribution of the tight-junction associated protein ZO-1 in the mouse testis. Am J Anat 191:35–47

Byers SW, Sujarit S, Jégou B, Butz S, Hoschutzky H, Herrenknecht K, MacCalman C, Blaschuk OW (1994) Cadherins and cadherin-associated molecules in the developing and maturing rat testis. Endocrinology 134:630–639

Cameron DF, Muffly KE (1991) Hormonal regulation of spermatid binding. J Cell Sci 100:623–633

Cancilla B, Risbridger GP (1998) Differential localization of fibroblast growth factor receptor-1, -2, -3, and –4 in fetal, immature, and adult rat testes. Biol Reprod 58:1138–1145

Carreau S, Genissel C, Bilinska B, Levallet J (1999) Sources of oestrogen in the testis and reproductive tract of the male. Int J Androl 22:211–223

Castellon EA (1999) Influence of age, hormones, and germ cells on glutathione S-transferase activity in cultured Sertoli cells. Int J Androl 22:49–55

Chaudhary J, Skinner MK (1999) E-box and cyclic adenosine monophosphate response elements are both required for follicle stimulating hormone-induced transferrin promoter activation in Sertoli cells. Endocrinology 140:1262–1271

Chen Y, Dicou E, Djakiew D (1997) Characterization of nerve growth factor precursor protein expression in rat round spermatids and the trophic effects of nerve growth factor in the maintenance of Sertoli cell viability. Mol Cell Endocrinol 127:129–136

Cheng CY, Bardin CW (1986) Rat testicular testibumin is a protein responsive to follicle stimulating hormone and testosterone that shares immunodeterminants with albumin. Biochemistry 25:5276–5288

Cheng CY, Grima J, Stahler MS, Guglielmotti A, Silvestrini B, Bardin CW (1990) Sertoli cell synthesizes and secretes a protease inhibitor, a_2-macroglobulin. Biochemistry 29:1063–1068

Cheng CY, Grima J, Stahler MS, Lockshin RA, Bardin CW (1989) Testins are structurally related Sertoli cell proteins whose secretion is tightly coupled to the presence of germ cells. J Biol Chem 264:21386–21393

Cheng CY, Mathur PP, Grima J (1988) Structural analysis of clusterin and its subunits in ram rete testis fluid. Biochemistry 27:4079–4088

Cheng CY, Morris PL, Mathur PP, Marshall A, Grima J, Turner TT, Carreau S, Tung KSK, Bardin CW (1988) Identification of a factor in fractions of Sertoli cell enriched culture medium containing CMB-21 that stimulates androgen production by Leydig cells. In: Nonsteroidal Gonadal Factors: Physiological Roles and Possibilities in Contraceptive Development. Eds. Hodgen G, Rosenwachs Z, Speiler J. Norfolk, Jones Institute Press, pp. 192–201

Cheng CY, Morris PL, Zwain I, Carreau S, Drosdowsky MA, Bardin CW (1993) A Leydig cell stimulator from Sertoli cells. In: Sexual Precocity. Etiology, Diagnosis, and Management. Eds. Grave GD, Cutler GB. New York, Raven Press, pp. 221–227

Chung SSW, Lee WM, Cheng CY (1999b) A study on the formation of specialized inter-Sertoli cell junctions in vitro. J Cell Physiol 181:258–272

Chung SSW, Mo MY, Silvestrini B, Lee WM, Cheng CY (1998a) Rat testicular N-cadherin: its complementary deoxyribonucleic acid cloning and regulation. Endocrinology 139:1853–1862

Chung SSW, Mruk D, Lee WM, Cheng CY (1999a) Identification and purification of proteins from germ cell-conditioned medium. Biochem Mol Biol Int 47:479–491

Chung SSW, Zhu L-J, Mo MY, Silvestrini B, Lee WM, Cheng CY (1998b) Evidence for cross-talk between Sertoli and germ cells using selected cathepsins as markers. J Androl 19:686–703

Collard MW, Day R, Akil H, Uhler MD, Douglass JO (1990) Sertoli cells are the primary site of prodynorphin gene expression in rat testis: regulation of mRNA and secreted peptide levels by cyclic adenosine 3',5'-monophosphate analogs in cultured cells. Mol Endocrinol Metab 72:1332–1339

Cunningham DB, Segretain D, Arnaud D, Rogner UC, Avner P (1998) The mouse Tsx gene is expressed in Sertoli cells of the adult testis and transiently in premeiotic germ cells during puberty. Dev Biol 15:345–360

Cyr DG, Hermo L, Egenberger N, Mertineit C, Trasler JM, Laird DW (1999) Cellular immunolocalization of occludin during embryonic and postnatal development of the mouse testis and epididymis. Endocrinology 140:3815–3825

Davidoff MS, Middendorff R, Pusch W, Muller D, Wichers S, Holstein AF (1999) Sertoli and Leydig cells of the human testis express neurofilament triplet proteins. Histochem Cell Biol 111:173–187

De Cesaris P, Starace D, Ricciolo A, Padula F, Filippini A, Ziparo E (1998) Tumor necrosis factor-α induces interleukin-6 production and integrin expression by distinct transduction pathways. J Biol Chem 273:7566–7571

Dejucq N, Dugast I, Ruffault A, Van der Meide PH, Jégou B (1995) Interferon-α and -γ expression in the rat testis. Endocrinology 136:4925–4931

Dejucq N, Lienard MO, Guillaume E, Dorval I, Jégou B (1998) Expression of interferons-α and -γ in testicular interstitial tissue and spermatogonia of the rat. Endocrinology 139:3081–3087

de Kretser DM, Kerr JB (1988) The cytology of the testis. In: The Physiology of Reproduction. Vol 1. Eds. Knobil E and Neill J. New York, Raven Press, pp. 837–932

Delfino F, Walker WH (1998) Stage-specific nuclear expression of NF-β in mammalian testis. Mol Endocrinol 12:1696–1707

Dorrington JH, Armstrong DT (1975) Follicle-stimulating hormone stimulates estradiol-17β synthesis in cultured Sertoli cells. Proc Natl Acad Sci USA 72:2677–2681

Dutertre M, Rey R, Porteu A, Josso N, Picard JY (1997) A mouse Sertoli cell line expressing anti-Mullerian hormone and its type II receptor. Mol Cell Endocrinol 136:57–65

Dym M, Lamsam-Casalotti S, Jia MC, Kelinman HK, Padadopoulos V (1991) Basement membrane increases G-protein levels and follicle stimulating hormone responsiveness of Sertoli cell adenylyl cyclase activity. Endocrinology 128:1167–1176

Eikvar L, Levy FO, Attramadal H, Jutte NH, Froysa A, Tvermyr SM, Hansson V (1985) Glucagon-stimulated cyclic AMP production and formation of estradiol in Sertoli cell cultures from immature rats. Mol Cell Endocrinol 39:107–113

Enders GC, Millette CF (1988) Pachytene spermatocyte and round spermatid binding to Sertoli cells in vitro. J Cell Sci 90:105–114

Ergun S, Harneit S, Paust HJ, Mukhopadhyay AK, Holstein AF (1998) Endothelin and endothelin receptors A and B in the human testis. Anat Embryol (Berl) 199:207–214

Eskild W, Ree AH, Levy FO, Jahnsen T, Hansson V (1991) Cellular localization of mRNAs for retinoic acid receptor-α, cellular retinol-binding protein, and cellular retinoic acid-binding protein in rat testis: evidence for germ cell-specific mRNA. Biol Reprod 44:53–61

Esnard A, Esnard F, Guillou F, Gauthier F (1992) Production of the cysteine proteinase inhibitor cystatin C by rat Sertoli cells. FEBS Lett 300:131–135

Feig LA, Bellve AR, Erickson NH, Klagsbrun M (1980) Mitogenic polypeptide of the mammalian seminiferous epithelium: biology of the seminiferous growth factor (SGF). Proc Natl Acad Sci USA 77:4774–4778

Feig LA, Klagsbrun M, Bellve AR (1983) Sertoli cells contain a mitogenic polypeptide. J Cell Biol 97:1435–1443

Frayne J, Nicholson HD (1998) Localization of oxytocin receptors in the human and macaque monkey male reproductive tracts: evidence for a physiological role of oxytocin in the male. Mol Hum Reprod 4:527–532

Fujisawa M, Bardin CW, Morris PL (1992) A germ cell factor(s) modulates preproenkephalin gene expression in rat Sertoli cells. Mol Cell Endocrinol 84:79–88

Gaemers IC, Van Pelt AM, Van der Saag PT, Hoogerbrugge JW, Themmen AP, De Rooij DG (1998) Differential expression pattern of retinoid X receptors in adult murine testicular cells implies varying roles for these receptors in spermatogenesis. Biol Reprod 58:1351–1356

Gérard N, Corlu A, Kneip B, Kercret H, Rissel M, Guguen-Guillouzo C, Jégou B (1995) Liver-regulating protein (LRP) is a plasma membrane protein involved in cell contact mediated regulation of Sertoli cell function by primary spermatocytes. J Cell Sci 108:917–925

Grima J, Calcagno K, Cheng CY (1996) Purification, cDNA cloning, and developmental stages in the steady-state mRNA level of rat testicular tissue inhibitor of metalloproteases-2 (TIMP-2). J Androl 17:263–275

Grima J, Wong CCS, Zhu LJ, Zong SD, Cheng CY (1998) Testin secreted by Sertoli cells is associated with the cell surface, and its expression correlates with the disruption of Sertoli-germ cell junctions but not the inter-Sertoli cell junction. J Biol Chem 273:21040–21053

Griswold MD (1993) Protein secretion by Sertoli cells: general considerations. In: The Sertoli Cell. Eds. Russell LD, Griswold MD. Clearwater, Cache River Press, pp 195–200

Griswold MD (1995) Interactions between germ cells and Sertoli cells in the testis. Biol Reprod 52:211–216

Guillaumot P, Benahmed M (1999) Prolactin receptors are expressed and hormonally regulated in rat Sertoli cells. Mol Cell Endocrinol 149:163–168

Gustafsson K, Söder O, Pollanen P, Ritzén EM (1985) Isolation and partial characterization of an interleukin-1-like factor from rat testis interstitial fluid. J Reprod Immunol 14:139–150

Guttman JA, Mulholland DJ, Vogl AW (1999) Plectin is a concentrated at intercellular junctions and at the nuclear surface in morphologically differentiated rat Sertoli cells. Anat Rec 254:418–428

Hadley MA, Weeks BS, Kleinman HK, Dym M (1990) Laminin promotes formation of cord-like structures by Sertoli cells in vitro. Dev Biol 140:318–327

Hagenäs L, Ritzén EM, Ploön L, Hansson V, French FS, Nayfeh SN (1975) Sertoli cell origin of testicular androgen binding protein (ABP). Mol Cell Endocrinol 2:339–350

Hakovirta H, Yan W, Kaleva M, Zhang F, Vanttinen K, Morris PL, Söder M, Parvinen M, Toppari J. (1999) Function of stem cell factor as a survival factor of spermatogonia and localization of messenger ribonucleic acid in the rat seminiferous epithelium. Endocrinology 140:1492–1498

Hall PF (1988) Testicular steroid synthesis: organization and regulation. In: The Physiology of Reproduction. Vol 1. Eds. Knobil E, Neil JD. New York, Raven Press, pp 975–998

Hall SH, Conti M, French FS, Joseph DR (1990) Follicle-stimulating hormone regulation of androgen-binding protein messenger RNA in Sertoli cell cultures. Mol Endocrinol 4:349–355

Howe CC, Overton GC, Sawicki J, Solter D, Stein P, Strickland S (1988) Expression of SPARC/osteonectin transcript in murine embryos and gonads. Differentiation 37:20–25

Huggenvik J, Griswold MD (1981) Retinol binding protein in rat testicular cells. J Reprod Fertil 61:403–408

Igdoura SA, Rasky A, Morales CR (1996) Trafficking of sulfated glycoprotein-1 (prosaposin) to lysosomes or to the extracellular space in rat Sertoli cells. Cell Tissue Res 283:385–394

Ito E, Toki T, Ishihara H, Ohtani H, Gu L, Yokoyama M, Engel JD, Yamamoto M (1993) Erythroid transcription factor GATA-1 is abundantly transcribed in mouse testis. Nature 362:466–468

Ivell R, Balvers M, Rust W, Bathgate R, Einspanier A (1997) Oxytocin and male reproductive function. Adv Exp Med Biol 424:253–264

Jaillard C, Chatelain PG, Saez JM (1987) In vitro regulation of pig Sertoli cell growth and function: effects of fibroblast growth factors and somatomedin-c. Biol Reprod 37:665–674

Jégou B (1993) The Sertoli-germ cell communication network in mammals. Int Rev Cytol 147:25–96

Jégou B, Sharpe RM (1993) Paracrine mechanisms in testicular control. In: Molecular Biology of the Male Reproductive System. Eds. de Kretser DM. New York, Academic Press, pp 274–322

Jenab S, Morris PL (1996) Differential activation of signal transducer and activator of transcription (STAT)-3 and STAT-1 transcription factors and c-fos messenger ribonucleic acid by interleukin-6 and interferon-γ in Sertoli cells. Endocrinology 137:4738–4743

Jenab S, Morris PL (1997) Transcriptional regulation of Sertoli cell immediate early genes by interleukin-6 and interferon-g is mediated through phosphorylation of STAT-3 and STAT-1 proteins. Endocrinology 138:2740–2746

Jenab S, Morris PL (1998) Testicular leukemia inhibitory factor (LIF) and LIF receptor mediate phosphorylation of signal transducers and activators of transcription (STAT)-3 and STAT-1 and induce c-fos transcription and activator protein-1 activation in rat Sertoli but not germ cells. Endocrinology 139:1883–1890.

Jiang C, Hall SJ, Boekelheide K (1997) Cloning and characterization of the 5' flanking region of the stem cell factor gene in rat Sertoli cells. Gene 185:285–290

Jonsson CK, Zetterstrom RH, Holst M, Parvinen M, Söder O (1999) Constitutive expression of interleukin-1a messenger ribonucleic acid in rat Sertoli cells is dependent upon interaction with germ cells. Endocrinology 140:3755–3761

Kaipia A, Penttila TL, Shimasaki S, Ling N, Parvinen M, Toppari J (1992) Expression of inhibin β_A and β_B, follistatin, and activin A receptor messenger ribonucleic acids in the rat seminiferous epithelium. Endocrinology 131:2703–2710

Kanzaki M, Fujisawa M, Okuda Y, Okada H, Arakawa S, Kamidono S (1996) Expression and regulation of neuropeptide Y messenger ribonucleic acid in cultured immature rat Leydig and Sertoli cells. Endocrinology 137:1249–1257

Kanzaki M, Morris PL (1998) Identification and regulation of testicular interferon-γ (INF-γ) receptor subunits: INF-γ enhances interferon regulatory factor-1 and interleukin-1β converting enzyme expression. Endocrinology 139:2636–2644

Kent J, Wheatley SC, Andrews JE, Sinclair AH, Koopman P (1996) A male-specific role for SOX9 in vertebrate sex determination. Development 122:2813–2822

Keon BH, Schafer S, Kuhn C, Grund C, Franke WW (1996) Symplekin, a novel type of tight junction plaque protein. J Cell Biol 134:1003–1018

Khan SA, Mirsafian M, Howdeshell K, Dorrington JH (1999) Transforming growth factor-beta inhibits DNA synthesis in immature rat Leydig cells in vitro. Mol Cell Endocrinol 148:21–28

Khan SA, Schmidt K, Hallin P, DiPauli R, DeGeyter C, Nieschlag E (1988) Human testis cytosol and ovarian follicular fluid contain high amounts of interleukin-1-like factor(s). Mol Cell Endocrinol 58:221–230

Kierszenbaum AL, Feldman M, Lea O, Spruill WA, Tres LL, Petrusz P, French FS (1980) Localization of androgen-binding protein in proliferating Sertoli cells in culture. Proc Natl Acad Sci USA 77:5322–5326

Kim KH, Wright WW (1997) A comparison of the effects of testicular maturation and aging on the stage-specific expression of CP-2/cathepsin L messenger ribonucleic acid by Sertoli cells of the Brown Norway rat. Biol Reprod 57:167–1477

Kissell K, Hamm S, Schulz M, Vecchi A, Garlanda C, Engelhardt B (1998) Immunohistochemical localization of the murine transferrin receptor (TfR) on blood-tissue barriers using a novel anti-TfR monoclonal antibody. Histochem Cell Biol 110:63–72

Klaij IA, Timmerman MA, Blok LJ, Grootegoed JA, De Jong FH (1992) Regulation of inhibin-β_B subunit mRNA expression in rat Sertoli cells: consequences for the production of bioactive and immunoreactive inhibin. Mol Cell Endocrinol 85:237–246

Klaij IA, Tosbosch AM, Themmen AP, Shimasaki S, De Jong FH, Grootegoed JA (1990) Regulation of inhibin-α and -β B subunit mRNA levels in rat Sertoli cells. Mol Cell Endocrinol 68:45–52

Lacroix M, Fritz IB. (1982) The control of the synthesis and secretion of plasminogen activator by rat Sertoli cells in culture. Mol Cell Endocrinol 26:247–258

Lamb DJ (1993) Growth factors and testicular development. J Urol 150:583–592

Lamb DJ, Spotts GS, Shubhada S, Baker KR (1991) Partial characterization of a unique mitogenic activity secreted by rat Sertoli cells. Mol Cell Endocrinol 79:1–12

Lee J, Richburg JH, Shipp EB, Meistrich ML, Boekelheide K (1999) The Fas system, a regulator of testicular germ cell apoptosis, is differentially upregulated in Sertoli cell versus gem cell injury of the testis. Endocrinology 140:852–858

Le Magueresse-Battistoni B, Morera AM, Goddard I, Benahmed M (1995) Expression of mRNAs for transforming growth factor-β receptors in the rat testis. Endocrinology 136:2788–2791

Le Magueresse-Battistoni B, Pernod G, Sigillo F, Kolodie L, Benahmed M (1998) Plasminogen activator inhibitor-1 is expressed in cultured rat Sertoli cells. Biol Reprod 59:591–598

Le Magueresse-Battistoni B, Wolff J, Morera AM, Benahmed M (1994) Fibroblast growth factor receptor type 1 expression during rat testicular development and its regulation in cultured Sertoli cells. Endocrinology 135:2404–2411

Li LY, Seddon AP, Meister A, Risley MS (1989) Spermatogenic cell-somatic cell interactions are required for maintenance of spermatogenic cell glutathione. Biol Reprod 40:317–331

Lim K, Hwang BD (1995) Follicle stimulating hormone transiently induces expression of proto-oncogene c-myc expression in primary Sertoli cell cultures from early pubertal and prepubertal rat. Mol Cell Endocrinol 28:51–56

Lim K, Yoo JH, Kim KY, Kweon GR, Kwak ST, Hwang BD (1994) Testosterone regulation of proto-oncogene c-myc expression in primary Sertoli cell cultures from prepubertal rats. J Androl 15:543–550

Lin ML, DePhillip RM (1996) Sex-dependent expression of placental (P)-cadherin during mouse gonadogenesis. Anat Rec 246:535–544

Lindsey JS, Wilkinson MF (1996) PEM: a testosterone- and LH regulated homeobox gene expressed in mouse Sertoli cells and epididymis. Dev Biol 179:471–484

Loganzo F Jr, Fletcher PW (1992) Follicle stimulating hormone increases guanine nucleotide-binding regulatory protein subunit α_i-3 mRNA but decreases α_i-1 and α_i-2 mRNA in Sertoli cells. Mol Endocrinol 6:1259–1267

Madhubala R, Shubhada S, Steinberger A, Tsai YH (1987) Inhibition of ornithine decarboxylase activity by follicle stimulating hormone in primary culture of rat Sertoli cells. J Androl 8:383–387

Maeda S, Nam SY, Fujisawa M, Nakamuta N, Ogawa K, Kurohmaru M, Hayashi Y (1998) Localization of cytoplasmic dynein light-intermediate chain mRNA in the rat testis using in situ hybridization. Cell Struct Funct 23:9–15

Maguire SM, Tribley WA, Griswold MD (1997) Follicle stimulating hormone (FSH) regulates the expression of FSH receptor messenger ribonucleic acid in cultured Sertoli cells and in hypophysectomized rat testis. Biol Reprod 56:1106–1111

Mathur PP, Grima J, Mo MY, Zhu LJ, Aravindan GR, Calcagno K, O'Bryan M, Chung S, Mruk D, Lee WM, Silvestrini B, Cheng CY (1997) Differential expression of multiple cathepsin mRNAs in the rat testis during maturation and following lonidamine induced tissue restructuring. Biochem Mol Biol Int 42:217–233

McClure RF, Heppelmann CJ, Paya CV (1999) Constitutive Fas ligand gene transcription in Sertoli cells is regulated by Sp1. J Biol Chem 274:7756–7762

Mead R, Hinchliffe SJ, Morgan BP (1999) Molecular cloning, expression and characterization of the rat analogue of human membrane cofactor protein. Immunology 98:137–143

Means AR, Huckins C (1974) Coupled events in the early biochemical actions of FSH on the Sertoli cells of the testis. Curr Top Mol Endocrinol 1:145–165

Miller MG, Mulholland DJ, Vogl AW (1999) Rat testis motor proteins associated with spermatid translocation (dynein) and spermatid flagella (kinesin II). Biol Reprod 60:1047–1056

Mita M, Hall PF (1982) Metabolism of round spermatids from rats: lactate as the preferred substrate. Biol Reprod 26:445–455

Morais da Silva S, Hacker A, Harley V, Goodfellow P, Swain A, Lovell-Badge R (1996) SOX9 expression during gonadal development implies a conserved role for the gene in testis differentiation in mammals and birds. Nat Genet 14:62–68

Morena AR, Boitani C, De Grossi S, Stefanini M, Conti M (1995) Stage and cell-specific expression of the adenosine 3',5'-monophosphate-phosphodiesterase genes in the rat seminiferous epithelium. Endocrinology 136:687–695

Morita K, Sasaki H, Fujimoto K, Furuse M, Tsukita S (1999) Claudin 11/OSP-based tight junctions of myelin sheaths in brain and Sertoli cells in testis. J Cell Biol 145:579–588

Mruk D, Cheng CY (1999) Sertolin is a novel testicular gene in studying Sertoli-germ cell interactions in the rat testis. J Biol Chem 274:27056–27068

Mruk D, Cheng CH, Cheng YH, Mo MY, Grima J, Silvestrini B, Lee WM, Cheng CY (1998) Rat testicular extracellular superoxide dismutase: its purification, cellular distribution, and regulation. Biol Reprod 59:298–308

Mruk D, Zhu LJ, Silvestrini B, Lee WM, Cheng CY (1997) Interactions of proteases and protease inhibitors in Sertoli-germ cell cocultures preceding the formation of specialized Sertoli-germ cell junctions in vitro. J Androl 8:612–622.

Mullaney BP, Skinner MK (1992) Basic fibroblast growth factor (bFGF) gene expression and protein production during pubertal development of the seminiferous tubule: follicle stimulating hormone-induced Sertoli cell bFGF expression. Endocrinology 131:2928–2934

Mullaney BP, Skinner MK (1993) Transforming growth factor-β (β1, β2, β3) gene expression and action during pubertal development of the seminiferous tubules: potential role at the onset of spermatogenesis. Mol Endocrinol 7:67–76

Muller D, Paust HJ, Middendorff R, Davidoff MS (1997) Nerve growth factor (NGF) receptors in male reproductive organs. Adv Exp Med Biol 424:157–158

Nargolwalla C, McCabe D, Fritz IB (1990) Modulation of levels of messenger RNA for tissue-type plasminogen activator in rat Sertoli cells, and levels of messenger RNA for plasminogen activator inhibitor in testis peritubular cells. Mol Cell Endocrinol 70:73–80

O'Bryan MK, Cheng CY (1997) Purification and cellular localization of β_2-microglobulin in the testis. Life Sci 61:487–494

O'Bryan MK, Grima J, Mruk D, Cheng CY (1997) Haptoglobin is a Sertoli cell product in the rat seminiferous epithelium: its purification and regulation. J Androl 18:637–645

Olaso R, Pairault C, Habert R (1998) Expression of type I and II receptors for transforming growth factor -β in the adult rat testis. Histochem Cell Biol 110:613–618

Ong DE, Chytil F (1978) Cellular retinoic acid-binding protein from rat testis. J Biol Chem 253:4551–4554

Orth JM, Christensen AK (1979) Localization of ^{125}I-labelled FSH in the testes of hypophysectomized rats by autoradiography at the light and electron microscope levels. Endocrinology 101:262–278

Oyen O, Sandberg M, Levy FO, Tasken K, Beebe S, Hansson V, Jahnsen T (1988) Molecular cloning and cell-specific expression of newly discovered subunits of cAMP-dependent protein kinases: implications for different cellular responses to cAMP. APMIS Suppl 2:238–250

Papadopoulos V (1991) Identification and purification of a human Sertoli cell-secreted protein (hSCSP-80) stimulating Leydig cell steroid biosynthesis. J Clin Endocrinol Metab 72:1332–1339

Papadopoulos V, Jia MC, Culty M, Hall PF, Dym M (1993) Rat Sertoli cell aromatase cytochrome P_{450}: regulation by cell culture conditions and relationship to the state of cell differentiation. In Vitro Cell Dev Biol Anim 29:943–949

Palombi F, Farini D, Salanova M, De Grossi S, Stefanini M (1992) Development and cyto-differentiation of peritubular myoid cells in rat testis. Anat Rec 233:32–40

Pelletier RM (1995) The distribution of connexin 43 is associated with the germ cell differentiation and with the modulation of the Sertoli cell junctional barrier in continual (guinea pig) and seasonal breeders' (mink) testes. J Androl 16:400–409

Pelletier RM, Okawara Y, Vitale ML, Anderson JM (1997) Differential distribution of the tight-junction-associated protein ZO-1 isoforms α^+ and α^- in guinea pig Sertoli cells: a possible association with F-actin and G-actin. Biol Reprod 57:367–376

Pelletier R, Trifaro JM, Carbajal ME, Okawara Y, Vitale ML (1999) Calcium-dependent actin-filament-severing protein scinderin levels and localization in bovine testis, epididymis, and spermatozoa. Biol Reprod 60:1128–1136

Persson H, Ayer-Lievre C, Söder O, Villar MJ, Metsis M, Olson L, Ritzén M, Hokfelt T (1990) Expression of β-nerve growth factor receptor mRNA in Sertoli cells down-regulated by testosterone. Science 247:704–707

Phamantu NT, Bonnamy PJ, Bouakka M, Bocquet J (1995) Inhibition of proteoglycan synthesis induces an increase in follicle stimulating hormone (FSH)-stimulated estradiol production by immature rat Sertoli cells. Mol Cell Endocrinol 109:37–45

Pineau C, Sharpe RM, Saunders PT, Gérard N, Jégou B (1990) Regulation of Sertoli cell inhibin production and of inhibin-α subunit mRNA levels by specific germ cell types. Mol Cell Endocrinol 72:13–22

Pollack SE, Furth EE, Kallen CB, Arakane F, Kiriakidou M, Kozarsky KF, Strauss JF III (1997) Localization of the steroidogenic acute regulatory protein in human tissues. J Clin Endocrinol Metab 82:4243–4251

Pöllanen P, Söder O, Parvinen M (1989) Interleukin-1-alpha stimulation of spermatogonial proliferation in vivo. Reprod Fertil Develop 1:85–87

Ratcliffe MJ, Rubin LL, Staddon JM (1997) Dephosphorylation of the cadherin-associated p100/p120 proteins in response to activation of protein kinase C in epithelial cells. J Biol Chem 272:1–8

Reader SC, Foster PM (1990) The in vitro effects of four isomers of dinitrotoluene on rat Sertoli and Sertoli-germ cell cocultures: germ cell detachment and lactate and pyruvate production. Toxicol Appl Pharmacol 106:287–294

Ree AH, Hansson V, Walaas SI, Eskild W, Tasken KA (1999) Calcium/phospholipid-dependent protein kinases in rat Sertoli cells: regulation of androgen receptor messenger ribonucleic acid. Biol Reprod 60:1257–1262

Rees JL, Daly AK, Redfern CPF (1989) Differential expression of the a and b retinoic acid receptors in tissues of the rat. Biochem J 259:917–919

Reynolds AB, Daniel J, McCrea P, Wheelock MJ, Wu J, Zhang Z (1994) Identification of a new catenin: The tyrosine kinase substrate $p120^{cas}$ associates with E-cadherin complexes. Mol Cell Biol 14:8333–8342

Rhee K, Wolgemuth DJ (1995) Cdk family genes are expressed not only in dividing but also in terminally differentiated mouse germ cells, suggesting their possible function during both cell division and differentiation. Dev Dyn 204:406–420

Risley MS, Tan IP, Roy C, Saez JC (1992) Cell-, age-, and stage-dependent distribution of connexin 43 gap junctions in testes. J Cell Sci 103:81–96

Rivarola MA, Sanchez P, Saez JM (1985) Stimulation of ribonucleic acid and deoxyribonucleic acid synthesis in spermatogenic cells by their coculture with Sertoli cells. Endocrinology 117:1796–1802

Roberts KP, Griswold MD (1990) Characterization of rat transferrin receptor cDNA: the regulation of transferrin receptor mRNA in testes and in Sertoli cells in culture. Mol Endocrinol 4:531–542

Rubin LL, Staddon JM (1999) The cell biology of the blood-brain barrier. Annu Rev Neurosci 22:11–28

Russell LD, Tallon DM, Weber JE, Wong V, Peterson RN (1983) Three-dimensional reconstruction of a rat stage V Sertoli cell: III. A study of specific cellular relationships. Am J Anat 167:181–192

Saitou M, Ando-Akatsuka Y, Itoh M, Furuse M, Inazawa J, Fujimoto K, Tsukita S (1997) Mammalian occludin in epithelial cells: its expression and subcellular distribution. Eur J Cell Biol 73:222–231

Salanova M, Ricci G, Boitani C, Stefanini M, De Grossi S, Palombi F (1998) Junctional contacts between Sertoli cells in normal and aspermatogenic rat seminiferous epithelium contain $\alpha 6\beta 1$ integrins, and their formation is controlled by follicle stimulating hormone. Biol Reprod 58:371–378

Salanova M, Stefanini M, De Curtis I, Palombi F (1995) Integrin receptor $\alpha 6\beta 1$ is localized at specific sites of cell-to-cell contact in rat seminiferous epithelium. Biol Reprod 52:79–87

Salas-Cortes L, Jaubert F, Barbaux S, Nessmann C, Bono MR, Fellous M, McElreavey K, Rosemblatt M (1999) The human SRY protein is present in fetal and adult Sertoli cells and germ cells. Int J Dev Biol 43:135–140

Samy ET, Li JCH, Grima J, Lee WM, Silvestrini B, Cheng CY (2000) Sertoli cell prostaglandin D2 synthetase (PGD-S) is a multi-functional molecule: its expression and regulation. Endocrinology 141:710–721

Sanchez A, Jimenez R, Burgos M, Diaz De la Guardia R (1994) A substance secreted by rat Sertoli cells indices feminization of embryonic chick testes in vitro. Anat Embryol (Berl) 189:531–537

Santiemma V, Salfi V, Casasanta N, Fabbrini A (1987) Lactate dehydrogenase and malate dehydrogenase of Sertoli cells in rats. Arch Androl 19:59–64

Saunders PT, Fisher JS, Sharpe RM, Millar MR (1998) Expression of oestrogen receptor-β (ER-β) occurs in multiple cell types, including some germ cells, in the rat testis. J Endocrinol 156:R13–17

Sharpe RM, Fraser HM, Cooper I, Rommerts FF (1981) Sertoli-Leydig cell communication via an LHRH-like factor. Nature 290:785–787

Shibamoto S, Hayakawa M, Takeuchi K, Hori T, Miyazawa K, Kitamura N, Johnson KR, Wheelock MJ, Matsuyoshi N, Takeichi, Ito F (1995) Association of p120, a tyrosine kinases substrate, with E-cadherin/catenin complexes. J Cell Biol 128: 949–957

Shibayama-Imazu T, Ogane K, Hasegawa Y, Nakajo S, Shioda S, Ochiai H, Nakai Y, Nakaya K (1998) Distribution of PNP-14 (β-synuclein) in neuroendocrine tissues: localization in Sertoli cells. Mol Reprod Dev 50:163–169

Shubhada S, Baker K, Baker S, Lamb DJ (1996) Sertoli cell secreted growth factor: cellular origin, endocrine, and paracrine regulation of secretion. J Androl 15:128–135

Shubhada S, Lin SN, Qian ZY, Steinberger A, Tsai YH (1989) Polyamine profiles in rat testis, germ cells, and Sertoli cells during testicular maturation. J Androl 10:145–151

Sigillo F, Guillou F, Fontaine I, Benahmed M, Le Magueresse-Battistoni B (1999) In vitro regulation of rat Sertoli cell transferrin expression by tumor necrosis factor-α and retinoic acid. Mol Cell Endocrinol 148:163–170

Skinner MK (1991) Cell-cell interactions in the testis. Endocr Rev 12:45–77

Skinner MK (1993) Secretion of growth factors and other regulatory factors. In: The Sertoli Cell. Eds. Russell LD and Griswold MD. Clearwater, Cache River Press, pp. 237–247

Skinner MK, Fritz IB (1985) Structural characterization of proteoglycans produced by testicular peritubular cells and Sertoli cells. J Biol Chem 260:11874–11883

Skinner MK, Griswold MD (1980) Sertoli cells synthesize and secrete transferrin-like protein. J Biol Chem 255:9523–9525

Skinner MK, Griswold MD (1983) Sertoli cells synthesize and secrete a ceruloplasmin-like protein. Biol Reprod 28:1225–1229

Skinner MK, Moses HL (1989) Transforming growth factor beta gene expression and action in the seminiferous tubule: peritubular cell-Sertoli cell interactions. Mol Endocrinol 3:625–634

Skinner MK, Takacs K, Coffey RJ (1989) Transforming growth factor-alpha gene expression and action in the seminiferous tubule: peritubular cell-Sertoli cell interactions. Endocrinology 124:845–854

Skinner MK, Tung PS, Fritz IB (1985) Cooperativity between Sertoli cells and testicular peritubular cells in the production and deposition of extracellular matrix components. J Cell Biol 100:1941–1947

Slegtenhorst-Eegdeman KE, Post M, Baarends WM, Themmen AP, Grootegoed JA (1995) Regulation of gene expression in Sertoli cells by follicle stimulating hormone (FSH): cloning and characterization of LRPR1, a primary response gene encoding a leucine-rich protein. Mol Cell Endocrinol 108:115–124

Slegtenhorst-Eegdeman KE, Verhoef-Post M, Parvinen M, Grootegoed JA, Themmen AP (1998) Differential regulation of leucine-rich primary response gene 1 (LRPR1) mRNA expression in rat testis and ovary. Mol Hum Reprod 4: 649–656

Smith EP, Hall SH, Monaco L, French FS, Wilson MW, Conti M (1989) A rat Sertoli cell factor similar to basic fibroblast growth factor increases c-fos messenger ribonucleic acid in cultured Sertoli cells. Mol Endocrinol 3:954–961

Sorrentino C, Silvestrini B, Braghiroli L, Chung SS, Giacomelli S. Leone MG, Xie Y, Sui Y, Mo M, Cheng CY (1998) Rat prostaglandin D2 synthetase: its tissue distribution, changes during maturation, and regulation in the testis and epididymis. Biol Reprod 59:843–853

Staddon JM, Herrenknecht K, Smales C, Rubin LL (1995) Evidence that tyrosine phosphorylation may increase tight junction permeability. J Cell Sci 108:609–619

Stahler MS, Schlegel C, Bardin CW, Silvestrini B, Cheng CY (1991) α_2-Macroglobulin is not an acute-phase protein in the rat testis. Endocrinology 128:2805–2814

Stephan JP, Syed V, Jégou B (1997) Regulation of Sertoli cell IL-1 and IL-6 production in vitro. Mol Cell Endocrinol 134:109–118

Suarez-Quian CA, Dai MZ, Onoda M, Kriss RM, Dym M (1989) Epidermal growth factor receptor localization in the rat and monkey testes. Biol Reprod 41:921–932

Subramanian S, Adiga PR (1996) Hormonal modulation of riboflavin carrier protein secretion by immature rat Sertoli cells in culture. Mol Cell Endocrinol 120:41–50

Subramanian S, Adiga PR (1999) Characterization and hormonal modulation of immunoreactive thiamin carrier protein in immature rat Sertoli cells in culture. J Steroid Biochem Mol Biol 68:23–30

Swift TA, Dias JA (1988) Testosterone suppression of ornithine decarboxylase activity in rat Sertoli cells. Endocrinology 123:687–693

Syed V, Söder O, Arver S, Lindh M, Khan S, Ritzen EM (1988) Ontogeny and cellular origin of an interleukin-1-like factor in the reproductive tract of the male rat. Int J Androl 11:437–447

Sylvester SR, Morales C, Oko R, Griswold MD (1991) Localization of sulfated glycoprotein-2 (clusterin) on spermatozoa and in the reproductive tract of the male rat. Biol Reprod 45:195–207

Sylvester SR, Skinner MK, Griswold MD (1984) A sulfated glycoprotein synthesized by Sertoli cells and by epididymal cells is a component of the sperm membrane. Biol Reprod 31:1087–1101

Tan IP, Roy C, Saez JC, Saez CG, Paul DL, Risley MS (1996) Regulated assembly of connexin 33 and connexin 43 into rat Sertoli cell gap junctions. Biol Reprod 54:1300–1310

Tohonen V, Osterlund C, Nordqvist K (1998) Testatin: a cystatin-related gene expressed during early testis development. Proc Natl Acad Sci USA 95:14208–14213

Tres LL, Smith EP, Van Wyk JJ, Kierszenbaum AL (1986) Immunoreactive sites and accumulation of somatomedin-C in rat Sertoli-spermatogenic cell co-cultures. Exp Cell Res 162:33–50

Tryggvason K (1993) The laminin family. Curr Opin Cell Biol 5:877–882

Tsai YN, Lin SN (1985) Differential release of polyamines by cultured rat Sertoli cells. J Androl 6:348–352

Tsuruta JK, O'Brien DA, Griswold MD (1993) Sertoli cell and germ cell cystatin C: stage-dependent expression of two distinct messenger ribonucleic acid transcripts in rat testis. Biol Reprod 49:1045–1054

Ueno N, Baird A, Esch F, Ling N, Guillemin R (1987) Isolation and partial characterization of basic fibroblast growth factor from bovine testis. Mol Cell Endocrinol 49:189–194

Ulisse S, Rucci N, Persanti D, Carosa E, Graziano FM, Pavan A, Ceddia P, Arizzi M, Muzi P, Cironi L, Gnessi L, D'Armiento M, Jannini EA (1998) Regulation by thyroid hormone of the expression of basement membrane components in rat prepubertal Sertoli cells. Endocrinology 139:741–747

van der Donk JA, de Ruiter-Bootsma AL, Ultee-van Gessel AM, Wauben-Penris PJ (1986) Cell-cell interaction between rat Sertoli cells and mouse germ cells in vitro. Expt Cell Res 164:191–198

Vannelli BG, Barni T, Orlando C, Natali A, Serio M, Balboni GC (1988) Insulin-like growth factor-I (IGF-I) and IGF-I receptor in human testis: an immunohistochemical study. Fertil Steril 49:666–669

Vigier B, Picard JY, Campargue J, Forest MG, Heyman Y, Josso N (1985) Secretion of anti-Mullerian hormone by immature bovine Sertoli cells in primary culture, studies by a competition-type radioimmunoassay: lack of modulation by either FSH or testosterone. Mol Cell Endocrinol 43:141–150

Wada M, Shintani Y, Kosaka M, Sano T, Hizawa K, Saito S (1996) Immunohistochemical localization of activin A and follistatin in human tissues. Endocr J 43:375–385

Waeber G, Meyer TE, LeSieur M, Hermann HL, Gérard N, Habener JF (1991) Developmental stage-specific expression of cyclic adenosine 3',5' monophosphate response element-binding protein CREB during spermatogenesis involves alternative exon splicing. Mol Endocrinol 5:1418–1430

Walker WH, Daniel PB, Habener JF (1998) Inducible cAMP early repressor ICER down-regulation of CREB gene expression in Sertoli cells. Mol Cell Endocrinol 143:167–178

Wang JE, Josefsen GM, Hansson V, Haugen TB (1998) Residual bodies and IL-1α stimulate expression of mRNA for IL-1α and IL-1 receptor type I in cultured rat Sertoli cells. Mol Cell Endocrinol 137:139–144

Weber JE, Russell LD, Wong V, Peterson RN (1983) Three-dimensional reconstruction of a rat stage V Sertoli cell: II. Morphometry of Sertoli-Sertoli and Sertoli-germ-cell relationships. Am J Anat 167:163–179

Weiner KX, Dias JA (1990) Protein synthesis is required for testosterone to decrease ornithine decarboxylase messenger RNA levels in rat Sertoli cells. Mol Endocrinol 4:1791–1798

Wells A, Ware MF, Allen FD, Lauffenburger DA (1999) Shaping up for shipping out: PLCγ signaling of morphology changes in EGF-stimulated fibroblast migration. Cell Mot Cytoskeleton 44:227–233

Welsh MJ, Gaestel M (1998) Small heat-shock protein family: function in health and disease. Ann NY Acad Sci 851:28–35

Welsh MJ, Wu W, Parvinen M, Gilmont RR (1996) Variation in expression of hsp27 messenger ribonucleic acid during the cycle of the seminiferous epithelium and co-localization of hsp27 and microfilaments in Sertoli cells of the rat. Biol Reprod 55:141–151

Wilkinson MF, Kleeman J, Richards J, MacLeod CL (1990) A novel oncofetal gene is expressed in a stage-specific manner in murine embryonic development. Dev Biol 141:451–455

Williams J, Foster PM (1988) The production of lactate and pyruvate as sensitive indices of altered rat Sertoli cell function in vitro following the addition of various testicular toxicants. Toxicol Appl Pharmacol 94:160–170

Wissenbach U, Schroth G, Philipp S, Flockerzi V (1998) Structure and mRNA expression of a bovine trp homologue related to mammalian trp2 transcripts. FEBS Lett 429:61–66

Wolf KW, Winking H (1996) The organization of microtubules and filamentous actin in cytospin preparations of Sertoli cells from w/wwei mutant mice devoid of germ cells. Anat Embryol (Berl) 193:413–417

Wong CCS, Chung SSW, Zhu LJ, Mruk D, Lee WM, Cheng CY (2000) Changes in the expression of junctional and nonjunctional complex component genes when inter-Sertoli tight junctions are formed in vitro. J Androl (in press)

Wong V, Russell LD (1983) Three-dimensional reconstruction of a rat stage V Sertoli cell: I. Methods, basic configuration, and dimensions. Am J Anat 167:143–161

Wright WW, Musto NA, Mather JP, Bardin CW (1981) Sertoli cells secrete both testis-specific and serum proteins. Proc Natl Acad Sci USA 78:7565–7569

Wrobel KH, Bickel D, Kujat R (1995) Distribution pattern of F-actin, vimentin, and α-tubulin in the bovine testis during postnatal development. Acta Anat (Basel) 153:263–272

Wrobel KH, Bickel D, Schimmel M, Kujat R (1996) Immunohistochemical demonstration of nerve growth factor receptor in bovine testis. Cell Tissue Res 285:189–197

Yan W, West A, Toppari J, Lahdetie J (1997) Stage-specific expression and phosphorylation of retinoblastoma protein (pRb) in the rat seminiferous epithelium. Mol Cell Endocrinol 132:137–148

Yomogida K, Ohtani H, Harigae H, Ito E, Nishimune Y, Engel JD, Yamamoto M (1994) Developmental stage- and spermatogenic cycle-specific expression of transcription factor GATA-1 in mouse Sertoli cells. Development 120:1759–1766

Yoshikawa K, Aizawa T (1988) Expression of the enkephalin precursor gene in rat Sertoli cells: regulation by follicle stimulating hormone. FEBS Lett 237:183–186

Zhai Y, Higgins D, Napoli JL (1997) Coexpression of the mRNAs encoding retinol dehydrogenase isoenzymes and cellular retinol-binding protein. J Cell Physiol 173 36–43

Zhu LJ, Krempels K, Bardin CW, O'Carroll AM, Mezey E (1998) The localization of messenger ribonucleic acids for somatostatin receptors 1, 2, and 3 in rat testis. Endocrinology 139:350–357

Zwain IH, Cheng CY (1994) Rat seminiferous tubular culture medium contains a biological factor that inhibits Leydig cell steroidogenesis: its purification and mechanism of action. Mol Cell Endocrinol 104:213–227

14 The Coming of Age of the Epididymis

B. Robaire, P. Syntin, K. Jervis

14.1 Historical Perspective: Development of an Awareness of the Epididymis As an Essential and Complex Tissue 229
14.2 Differentiation: the Formation of the Epididymal Duct 231
14.3 Post-Natal Development of the Epididymis: Multiple Cell Types with Different Functions 231
14.4 The Epididymis in the Adult: a Highly Specialized Tissue 234
14.5 The Aging Process: Changes in Structure and Functions of the Epididymis 248
References 251

14.1 Historical Perspective: Development of an Awareness of the Epididymis As an Essential and Complex Tissue

Several descriptive anatomical and histological studies of the testicular excurrent duct system from various species appeared at the beginning of the twentieth century. In the 1920's, Benoit published a series of articles which culminated in a comprehensive treatise (Benoit, 1926) on the anatomy, cytology, and histophysiology of the efferent ducts, epididymis, and vas deferens. In that article, we can recognize easily our current view that a series of efferent ducts merge to form a single, highly convoluted, epididymal duct that is made up of several cell types, is under the influence of testicular substances, and is the site where spermatozoa acquire their potential for motility. The prevalent hypothesis in the 1920's, that the epididymis plays an active role in the maturation of spermatozoa, their acquisition of motility, and their storage, was challenged by Young in a series of articles published between 1929 and 1931 (Young, 1929a,b, 1931). Relatively little research was done on the excurrent duct system during the ensuing three decades.

In the 1960's, a resurgence of interest in the epididymis was spearheaded by Orgebin-Crist's (Orgebin-Crist, 1967) and Bedford's (Bedford, 1967) demonstrations that spermatozoa did, in fact, mature, i.e. acquire the ability to swim and fertilize eggs by passing through this tissue. With the advent of sophisticated surgical techniques in the 1980's, it became possible to do vaso-epididymostomy, to obtain sperm from the proximal epididymis and the efferent ducts, and to use these sperm for in vitro fertilization; successful pregnancy led to a questioning of the importance of the epididymis in sperm maturation (Silber, 1989); however, the overwhelming weight of evidence clearly established that, in all mammals studied to date, sperm maturation must take place, under normal circumstances, in the epididymis (Cooper, 1990).

Since 1966, over 21,000 research articles have been published on the efferent ducts, epididymis, and vas deferens. The number of articles each year has steadily increased, from in the 100 range in the late 1960's to well over 600 in the 1990's. We have acquired an in depth understanding of some key facets of the structure, biochemical properties, and functions of this tissue, yet huge gaps remain in our understanding in all of these areas. The unforeseen complexity in the cellular properties, heterogeneity, spatial and temporal organization of protein synthesis and secretion, and dynamic interactions between the epithelial cell functions and the contents of the luminal compartment have presented new challenges. This complexity can be used to advantage in understanding basic biological processes of differentiation, development and aging, selective gene activation, and cell-cell interactions. There are strong reasons to believe that as our understanding of this tissue continues to grow, we will be able to target specific elements of this tissue to develop therapeutic agents to treat certain types of male infertility, to develop male contraceptive agents, and to gain an understanding of how environmental chemicals and stress conditions affect the functions of this key tissue.

14.2 Differentiation: the Formation of the Epididymal Duct

During embryonic development, mesonephric tubules give rise to the efferent ducts while the cranial pole of the Wolffian duct gives rise to the epididymis and the vas deferens (Byskov and Høyer, 1988). As gestation proceeds, the tubule becomes longer and the degree of convolution increases from the cranial to the distal pole of the epididymis. Although no apparent morphological differentiation of cell types is seen prior to birth, androgen receptors are found at least as early as gestational day 14 (rat) and increase during the remaining gestational period. Withdrawal of androgen during the last third of gestation results in regression of the epididymis. Based on the presence of several proteins with known function, e.g. aquaporin-1 with its role in fluid resorption, it would appear that epididymal epithelial cells have defined functions prior to birth (Fisher et al., 1998). Null mutation and tissue distribution studies have indicated that normal development of the epididymis during gestation is crucially dependent on the presence of several genes, including the growth factor Bmp8a (Zhao et al., 1998), the oncogene c-ros (Sonnenberg-Riethmacher et al., 1996), the paired box gene Pax-2 (Oefelein et al., 1996), and the homeobox domain related genes, Hox family of genes (Podlasek et al., 1999; Lindsey et al., 1996), and Pem2 (Sutton et al., 1998).

14.3 Post-Natal Development of the Epididymis: Multiple Cell Types with Different Functions

The postnatal development of the epididymis has been most extensively studied using the rat model. Few changes in epididymal histology occur during the first two weeks after birth of the rat pup (Sun and Flickinger, 1979). At the end of the second week, a population of differentiated cells of the immune system (halo cells) appears throughout the duct. Shortly thereafter, the undifferentiated epithelial cells give rise to columnar and narrow cells. The former differentiate into basal and principal cells at approximately four weeks after birth. It is important to note that, unlike in other tissues, basal cells are not precursor cells for principal cells (Clermont and Flannery, 1970). Approximately a week later, a subpopulation of principal cells, called apical cells, can be distinguished in

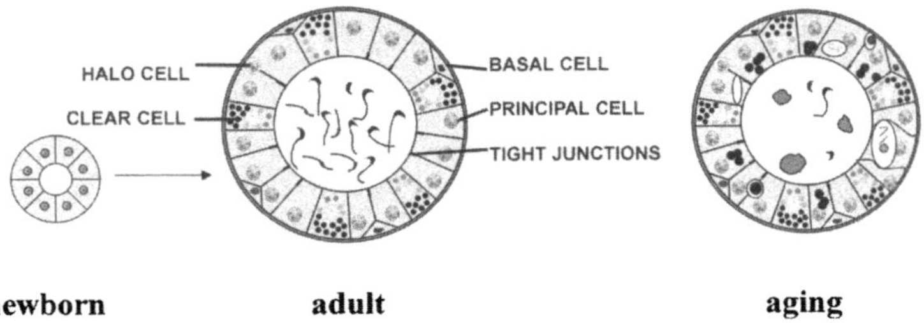

Fig. 1. Schematic representation of a cross section of a tubule of the caput epididymidis at birth and in the adult. Note the lack of differentiation of cell types at birth and the presence of several types of cells in the adult

the initial segment of the epididymis. At the same time, narrow cells beyond the initial segment begin to differentiate into clear cells; the appearance of these clear cells progresses posteriorly along the duct and is associated with the presence of spermatozoa (Hermo et al., 1992a).

As animals reach adulthood, the five major cell types found in the epididymal epithelium have acquired their characteristic appearance. These cell types are the principal, basal, narrow, clear, and halo cells (Fig. 1). Throughout the epididymis, the most abundant cell type found in the epididymal epithelium is the principal cell which has prominent stereocilia that extend into the lumen. In rats, principal cells constitute 80% of the total epithelial cell population in the initial segment, and this number gradually decreases to 65% in the cauda epididymidis (Robaire and Hermo, 1988). In the initial segment, principal cells are tall and columnar in shape. Ultrastructurally, the infranuclear region of principal cells is densely packed with rough endoplasmic reticulum; there is an abundant presence of large, dilated membranous elements in the apical region (Robaire and Hermo, 1988; Vierula et al., 1995). The supranuclear cytoplasm of principal cells is also characterized by the presence of large stacks of Golgi saccules, mitochondria, multivesicular bodies and smooth surfaced vesicles. Principal cells in the caput, corpus, and cauda epididymidis are relatively shorter in height, and have irregularly shaped nuclei. Abundant cisternae of rough endoplasmic reticulum, numerous mitochondria, and lipid droplets are seen in the infranuclear cytoplasm; a well-developed Golgi apparatus, lysosomes, multivesicular bodies, mitochondria, and endosomes, as well as polysomes, are present in the apical and supranuclear regions. At the luminal surface of principal cells, adluminal tight junctions exist between the principal cells and form the blood-epididymis barrier (Cyr et al., 1999; Levy and Robaire, 1999).

Basal cells are the second most abundant cell type found in the epididymal epithelium; they contribute about 10–20% of the total cells (Robaire and Hermo, 1988). The triangular, flat basal cells, found throughout the epididymis, reside near the base of the epithelium where they contact the basement membrane. Basal cells usually have elongated or round shaped nuclei and have projections to adjacent basal cells as well as between principal cells, toward the epididymal lumen (Veri et al, 1993). The exact functions of these cells are not known, however, some findings suggest that they may play a protective role in the epididymis.

Narrow cells, only found in the initial segment, are elongated cells between principal cells; their function remains unknown. In contrast, clear cells are found in the caput, corpus, and cauda epididymidis; they contribute 5–10% of total cells (Robaire and Hermo, 1988). Clear cells are so named because of their characteristic pale-staining cytoplasm at the level of the light microscope; they are characterized by the presence of numerous dense granules which are located above or below the nucleus, and a highly vacuolated apical region. The basal region, below the round and pale-stained nucleus, is full of pale or moderately dense bodies. Clear cells seem to play a major role in the uptake of particulate matter from the luminal compartment.

Halo cells are identified by their dark-stained nucleus surrounded by a pale-staining cytoplasm and are present at all levels of the epididymal epithelium. They constitute just under 5% of total epithelial cells. These cells originate from the immune system and are a combination of B and T lymphocytes as well as monocytes (Flickinger et al.,1997, Serre and Robaire, 1999).

In association with this morphological postnatal differentiation, the expression of a large number of biochemical markers changes as adulthood is attained. The specificity of epididymal secretion is progressively established with age in the rat (Ueda et al., 1990), mouse (Abou-Haila and Fain-Maurel, 1985), and boar (Syntin et al., 1999). The segment specific expression of a large number of genes and the checkerboard-like expression pattern among principal cells for many cell markers are striking hallmarks of epididymal differentiation.

14.4 The Epididymis in the Adult: a Highly Specialized Tissue

It is now well established that the epididymis is the site where spermatozoa mature (acquire the potential for motility and fertilization) as they traverse the proximal regions of the tissue and are then stored in the cauda epididymidis. In parallel with changes taking place in spermatozoa as they are moved along the epididymis, dynamic changes take place in the fluid bathing them. These changes are mediated by processes of absorption and secretion by epithelial cells lining the duct system. We will focus first on selected RNAs and/or proteins that show region or cell-specific patterns of expression in the epididymis. However, it is essential to dissociate between the RNAs and proteins that are present only in the epithelium from those proteins that are secreted apically toward the lumen of the tubule, and thus may be directly involved in the creation of the microenvironment that allows sperm to mature. We will then briefly discuss key elements of how the epididymis is regulated.

14.4.1 Epithelial Proteins

As noted above, the epididymis is a highly regionalized organ in its activities (absorption, secretion) and in its cellular composition. One of its most fascinating aspects is the complex and highly region specific pattern of gene expression and protein repartition throughout its length (Kirchoff, 1999). *In situ* hybridization and immunocytochemical analyses demonstrate the region-specific expression of many factors (Table 1). In addition to such region specific expressions, three other levels of complexity of gene expression in the epithelium of this tissue have become evident. These are cell type specific gene expression, a "checkerboard-like" pattern of gene expression for a given cell type within a specific region, and the different subcellular localizations of a given marker for a cell type in different regions of the tissue. Some examples will be provided to highlight this specificity of gene expression.

The regional variations among the different segments of the epididymis indicate that the epithelial cells along the duct have different functions and/or are under different regulatory control. This specific

Table 1. Regionalization of epididymal epithelial factors

Name	Species[1]	Regions[2] of highest expression	Suggested function	Regulation[3]	References
Polyoma virus enhancer activator 3, PEA3	R, M	IS	Transcription factors	A-raf	Lan et al, 99; Drevet et al, 98
Protooncogene A-*raf*	M	IS-Cap	2nd messenger signaling pathway	TF: βFGF	Winner & Wolgemuth, 95
Protooncogene c-ros	M	Cap	Receptor tyrosine kinase	ND	Sonnenberg-Riethmacher et al, 96
Pem homeobox	M, R	Cauda	Transcription factor	AR	Sutton et al, 98
Pax2	M	IS to Cda	Transcription factor	ND	Oefelein et al, 96
Bone morphogenic proteins, Bmp 8a, 7	M	IS	ND	ND	Zhao et al, 98; Chen et al., 99
Platelet-derived growth factor, PDGF	B	Cap	Growth factor	ND	Okamura et al, 95
Nerve growth factor βNGF	R, M	Corp	Growth factor	AR	Ayer-LeLievre et al, 88
Transforming growth factor, TGFβ3	R	Corp	Growth factor	ND	Desai et al, 98
5α-reductase 1	R	IS	Testo conversion	AR-TF	Viger & Robaire 96
5α-reductase 2	R	Cap	Testo conversion	AR-TF	Viger & Robaire, 96
Epididymal apical protein 1 EAP1	R, Mac	Cap	Desintegrin protease receptor	AR-TF	Perry et al, 92
Desintegrin & metallo-protease ADAM 7	M	Cap	Transmenbrane receptor	AR-TF	Cornwall et al, 97
Cadherin	R	Cap-Corp	Epithelium junctions	AR	Cyr et al, 93
Carbonic anhydrase II, IV	R	Corp	Acidification of epididymal fluid	AR	Kaunisto et al, 99

[1] Species: R-rat, M-mouse, B-Boar, Mac-Maccacque
[2] Region: IS-Initial segment, Cap-caput, corp-corpus, cda-cauda epidimymis
[3] AR-androgen-regulated, TF-regulated by testicular factor
ND-Not determined

expression should serve as a powerful tool to identify specific cells; it should now become possible to culture each cell type in order to determine their relative roles. For example, proteins found as cell surface receptors such as ADAM 7 (Cornwall and Hsia, 1997), low density lipoprotein receptor-related protein-2 (LRP-2, Hermo et al., 1999) or involved in the formation of the blood epididymis barrier such as E-cadherin (Cyr et al., 1992), or occludin (Cyr et al., 1999) could be used as markers for principal cells; glutathione S-transferase Yf (Veri et al., 1994), cyclo-oxygenase 1 (Cox 1, Wong et al., 1999), or superoxide dismutase (SOD, Perry et al., 1993a) could be used as markers for basal cells; α-mannosidase II (Igdoura and Hermo 1999) or aquaporin 9 (Andonian et al., 1999a) could be used as markers for clear cells.

The precise localization of lysosomal proteinases (cathepsins A, B and D) within the epididymal epithelium establishes that their expression is not only region but also cell-type specific. Clear cells of the caput epididymidis are intensely reactive for an anti-cathepsin D antibody, whereas those of the corpus and the cauda epididymides are unreactive (Igdoura et al., 1995; Andonian and Hermo 1999). Basal cells show a progressive increase in reactivity in the cauda epididymidis. Cathepsin A is present in principal, clear and basal cells, while cathepsin B is detected only in principal cells.

Immunocytochemical localization of six of the subunits of the glutathione S-transferase family of enzymes also showed characteristic regional variations in the staining of epithelial cells along the epididymis (Papp et al., 1995). One of the most striking changes is the highly selective staining of basal cells for the Yf subunit in the corpus and proximal cauda epididymidis, whereas in the initial segment and proximal caput epididymidis, both principal and basal cells were stained (Veri et al., 1993).

A characteristic checkerboard-like pattern can be seen for epididymal principal cells after immunolocalization of a number of gene products such as immobilin (Ruiz-Bravo, 1988, Hermo et al., 1992b), clusterin (SGP-2, Hermo et al., 1991), and mouse epididymal protein 9 (Rankin et al., 1992). Normally, either at the point in the duct where immunostaining first appears or when the signal stops being expressed there are, in a given cross section, some cells that are intensely stained while others remain unreactive. This functional mosaicism of cell types of the epithelium is not reflected by their morphology. The molecular mechanisms responsible for this mosaicism is elusive since adjacent epithelial cells are exposed to similar factors from the luminal compartment and the peripheral circulation. Possible explanations for this phenomenon are that individual cells or groups of cells have differing expression of receptors for factors that elicit specific responses, or that these cells are out of phase with one another.

In addition, some subcellular compartmental localization can be detected within a given cell type. For instance, cyclo-oxygenase 2 (cox-2) immunoreactivity is detected only in the apical pole of the principal cells (Wong et al., 1999). Different subcellular localizations can be seen at the microscopic level for different types of mannosidases; the α-mannosidase-Ib is localized mainly over cis-saccules while α-mannosidase-Ia is distributed mainly over trans-saccules of the Golgi stack. The distribution of steroid 5α-reductase is both peri-nuclear and supra-nuclear in the initial segment, but changes to primarily an apical localization further down the duct (Viger and Robaire, 1994).

14.4.2 Secreted Proteins

Sperm maturation is not intrinsic to the sperm cells themselves, but rather requires interaction with luminal epididymal proteins. The region-specific pattern of protein secretion is considered essential for the sequential multi-step process of sperm maturation. A better understanding of the regionalized gene expression or protein secretion is required to provide an accurate understanding of sperm maturation.

A considerable amount of information has become available recently regarding the proteins that are secreted by the epididymis (Table 2). These proteins have been studied using both *in vivo* and *in vitro* methods. *In vitro*, the incorporation of amino acids or sugars shows that the epididymal epithelium is active in the synthesis and secretion of glycoproteins (Brooks, 1983; Olson and Hinton, 1985; Holland and Orgebin-Crist, 1988; Turner et al., 1995; Syntin et al., 1996). Many of these proteins are secreted in a highly regionalized manner; it is rare to see any protein secreted all along the epididymis. The differences observed between *in vitro* and *in vivo* studies of protein secretion are likely attributable to the role played by the luminal presence of specific chemicals leading to a balance between secretion, reabsorption and degradation.

14.4.2.1 Regionalization of Secretion

In most of the species studied, the anterior part of the epididymis has been shown to be the most active in protein secretion; such species include the guinea-pig (Del-Rio, 1979), rat (Brooks, 1981), tammar Wallaby (Chaturapanich et al., 1992), and boar (Syntin et al., 1996). Some specific genes or proteins are expressed only in the initial segment of this tissue, e.g. proenkephalin (Garrett et al., 1990), CRES (Cornwall

Table 2. Regional secretion of epididymal proteins into the lumen

Name	Species[1]	kDa	Regions[2] of highest expression	Sperm associated	Suggested function	Regulation[3]	References
B/C-ERABP ESP1, MEP10	R, M	16–18	Cap-corp	Yes	Retinoid acid binding protein	AR	Brooks et al, 86; Lareyre et al, 99; Syntin et al, 96
D/E, AEG, CRISP-1, MEP7	R, M	27–37	Cap-corp-cda	Yes	Carboxypeptidase Y-like, Sperm egg fusion	AR	Cameo & Blaquier, 76 Brooks, 83b
Superoxide dismutase, SOD	R	ND	Cda	ND	Reduces reactive oxygen, Protection of sperm	AR	Perry et al, 93a
γ-Glutamyl trans-peptidase IV	R	71	IS, not cap	ND	Degrades GSSG, Protection of sperm	AR-TF	Palladino & Hinton, 94 Hinton et al, 95
Glutathione peroxidase GPX-5	Many, not Hu	23–25	IS-Cap	Yes	Reduces reactive oxygen Protection of sperm	AR-TF	Ghyselink et al, 90 Okamura et al, 97
Glutathione peroxidase GPX-3	M	23–25	Corp-cda	ND	Reduces reactive oxygen	AR	Schwaab et al, 98
Clusterin, SGP-2, apo J, SP40–40	Many	66–70	Cap	Yes	Complement lysis inhibitor	AR	Sylvester et al, 91; Cyr & Robaire, 92
Procathepsin L	R, B	39–40	Cap	ND	Intracellular recycling, Protease	local factor, PDGF	Tomomasa et al, 94 Okamura et al, 95a
Mannosidase	R, B	135	Cap-corp	Yes	Glycosidase, Sperm-egg fusion	AR	Okamura et al, 92, 95b Tulsiani et al, 95b Cornwall et al, 91
Hexosaminidase N-acetyl glucosaminidase	B	65	Cap	Yes	Glycosyltransferase, Modification of sperm	AR	Takada et al, 94 Syntin et al, 96 Miranda et al, 95

HE1, EP I-1, ESP14.6	Hu, many	19–21	Cap-corp	Yes	Cholesterol transferase protein, Modification of sperm	AR	Kirchhoff et al, 96 Okamura et al, 99
Human epidi-dymal protein HE2	Hu	ND	Cap-corp	Yes	Sperm-egg fusion	ND	Osterhoff et al, 94
Lactoferrin	M, B	75–85	Cap-corp	Yes	Protection of sperm	AR	Jin et al, 97; Yu et al, 93 Okamura et al, 97
P34H, P26H	Hu, Ha	34, 26	Corp	Yes	Dehydrogenase/reductase, Zona binding	ND	Legaré et al, 99
CRES	R, M	14, 19	IS	Yes	Proteinase inhibitor	TF	Cornwall et al, 92
HE4 Human epididymal prot	Hu	10	Corp-cda	ND	Extracellular proteinase inhibitor	ND	Kirchhoff et al, 91
HE5, cdw52 CAMPATH-1	Hu	26	Corp-cda	Yes	Decapacitation factor	AR	Kirchhoff et al, 93 Pera et al, 97
Pro-opiomela-nocortin POMC	R, M	26, 32	Caput	ND	Paracrine factor	ND	Jeannote et al, 90
Proenkephamin A	R, Ha	28	IS	Yes	ND	TF	Garrett et al, 90

[1] Species: R-rat, M-mouse, Ha- hamster, Hu-human, B-boar
[2] Region: IS -initial segment, Cap-caput, corp-corpus, cda-cauda epididymidis
[3] AR-androgen-regulated, TF-regulated by testicular factor
ND-not determined

et al., 1992), gamma-glutamyl-transpeptidase (Palladino and Hinton, 1994), glutathione peroxidase 5 (GPX5, Ghyselink et al., 1990), and the train A in the boar epididymis (Syntin et al., 1996). Although the functional identity of train A is still unknown, it is likely to be of major importance for several reasons; these include the fact that its secretion represents more than 75% of all secreted proteins in this segment, and that its electrophoretic characteristics on a 2D gel indicate that there are at least 6 isoforms ranging from 25 to 41 kDa with isoelectric points ranging from 5 to 6.6, presumably a result of high glycosylation or phosphorylation (Ravet et al. 1991; Dacheux et al., 1989). Some similar characteristics were observed for clusterin (Sylvester et al., 1984; Cheng et al., 1988; Mattmueller and Hinton, 1991), which is secreted mainly in the caput epididymidis, as well as for procathepsin (Okamura et al.,1995a), and some sugar oxidases (hexoseaminidase, mannosidase, Hall et al., 1996; Syntin et al., 1996; Okamura et al., 1995b; Tulsiani et al., 1993a).

Several proteins are secreted in both the caput and corpus epididymidis, for instance a retinol acid binding protein (RABP, equivalent to protein B/C, Ong et Chytil, 1988) and lactoferrin (Jin et al., 1997). In humans, lactoferrin is detected in the prostate and seminal vesicles, while the testis, epididymis and vas deferens do not seem to express this protein (Wichmann et al., 1989). Some proteins secreted more distally (corpus and cauda epididymidis) include HE4 (extracellular proteinase inhibitor, Kirchhoff et al., 1991), HE5 (cdw52, Pera et al., 1997) and GPX3 (Schwaab et al., 1998). Northern-blot analysis reveals that the transcript corresponding to superoxide dismutase is expressed principally in the cauda region of the epididymis; this is consistent with the high levels of superoxide dismutase enzyme activity found in cauda epididymal fluid (Perry et al., 1993a). It is interesting to note that very few proteins are secreted specifically in the cauda epididymidis. The reason for this is not clear, however, we can speculate that this occurs either because there is no need for new secretions, as the sperm maturation process is complete, or because proteins secreted in more proximal segments are accumulated in the fluid and remain present around the sperm within the cauda epididymidis. The lower height of principal cells and their less well developed Golgi complexes and endoplasmic reticulum in this region are consistent with this observation (Robaire and Hermo, 1988).

14.4.2.2 Putative Functions

Most of the secreted proteins that have been identified appear to be conserved among species and present in many mammalian species studied. However, the sequential order of their secretion may vary between species, and the importance of this order is far from being resolved.

During maturation, several modifications of the sperm surface can be attributed to exogenous enzyme activities such as glycosylation or proteolytic processing of sperm surface proteins (Tulsiani et al., 1998). At least four different enzymes involved in the addition of sugars (glycosyltransferases) are found: sialyltransferase, galactosyltransfrase, N-acetylglucosidase (hexosaminidase), and fucosyltransferase (Hamilton, 1980; Jauhiainen and Vanha-Pertulla, 1985; Tulsiani et al., 1993b). On the other hand, some glycosidases, enzymes that cleave sugars from glycoproteins are also found in the epididymal fluid; these include galactosidase (Tulsiani et al., 1995a) and mannosidase (Tulsiani et al., 1993a; Okamura et al., 1992). Proteolytic enzymes such as carboxypeptidase Y (protein D/E, Cameo and Blaquier, 1976), phosphatases (Mayorga and Bertini, 1985), and cathepsins (Tomomasa et al., 1994; Okamura et al., 1995a) have been characterized in the epididymis. The presence of powerful protease inhibitors (CRES, HE4) suggests that they may act to counterbalance the activity of endogenous proteases (Cornwall et al., 1995a; Perry et al., 1993b; Kirchhoff et al., 1991). Clearly, the relative amount and activities of the enzymes and of their specific inhibitors must provide the basis for the intricate balance necessary for remodeling sperm surface proteins.

High levels of several proteins that mediate antioxidant action are found in epithelial cells (4.1) and in the epididymal fluid (Hinton et al., 1995); they may protect epithelial and/or sperm cells against oxidative stress from exposure to endogenous reactive oxygen species or exogenous toxic substances. Several protective systems that are present in the luminal fluid include gamma-glutamyl transpeptidase (GGT, Lan et al., 1998), glutathione peroxidase (GPX, Ghyselink et al. 1990; Okamura et al., 1997) and superoxide dismutase (SOD, Perry et al., 1993a). Expression of the secretory antioxidant factors is region-specific, indicating that the need for antioxidant enzymes may vary along the epididymis. Some of the other major protein secreted by the epididymal epithelium possess carrier functions for the transport of retinol (RABP, Porter et al.,

1985) or of cholesterol (CTP, HE1, Kirchhoff et al.,1996; Baker et al.,1993; Okamura et al.,1999).

A number of glycoproteins on the surface of ejaculated spermatozoa are acquired from epithelial cell secretions during epididymal transit and have been described as potential candidates involved in mediating sperm-egg interaction. Whether any one protein is critical for this function has not yet been resolved. Of these glycoproteins, protein D/E, also know as acidic epididymal glycoprotein (AEG) is one that is extensively characterized (Lea and French, 1981; Brooks et al., 1986; Rochwerger et al., 1992); it belongs to the CRISP family (cysteine-rich secretory protein, Xu et al., 1997). Other candidate proteins include the α-mannosidase in mouse (Cornwall et al., 1991), and rat (Tulsiani et al., 1995b); the human protein 5 (HE5, also known as cdw52, Kirchhoff et al., 1993; Pera et al., 1997), and a protein of 34 kDa secreted in the human corpus epididymidis (Legaré et al., 1999), related to the p26 of hamster (Bérubé and Sullivan, 1994).

14.4.3 Regulation of Epididymal Functions

The dependence of the epididymis on androgens secreted into the peripheral circulation as well as on factors that are secreted directly into the epididymal lumen from the seminiferous tubules, rete testis and efferent ducts is well established (Blanchard and Robaire, 1997). It would appear logical that some epididymal specific or region-specific transactivation should regulate the complex gene expression seen along the epididymis, yet no such factors have yet been identified. However, several transcription factors including Bmp8a, (Zhao et al., 1998), PEA3 (Drevet et al., 1999; Lan et al, 1999), Pem-2 (Sutton et al., 1998) and Hox group (Podlasek et al., 1999; Lindsey et al., 1996) and proto-oncogenes c-ros (Sonnenberg-Riethmacher et al., 1996) and A-raf (Winer and Wolgemuth, 1995) show a region specific expression suggesting that they are good candidates for region-specific gene activation.

Even though testosterone and its metabolites are clearly the primary hormonal regulator of epididymal functions, there is strong evidence that other hormones play a role in modulating the functions of this tissue. There is substantial evidence that estradiol has a role in the epididymis. Estrogen effects on the epididymis can be viewed either as indirect or direct. Since estradiol is a potent suppressor of LH secretion, administration of estradiol results in a decrease in gonadotropins and consequently in a shutdown of Leydig cell function, and hence androgen production. However, there are several lines of evidence that suggest that estradiol has direct actions on the epididymis. These include the presence of aromatase activity on spermatozoa entering the epididymis (Janulis et al., 1998), and of estrogen receptors in epididymal principal cells (Hess et al., 1997), the consequences of null mutations of the estrogen receptors (Couse et al., 1997), and the sequella of treating with an estrogen receptor antagonist (Belmonte et al., 1998).

Other hormones for which either receptors or the hormone itself have been found in the epididymis include prolactin (Orgebin-Crist and Djiane, 1979) which seems to affect protein secretion in caput epididymidis, retinoic acid, for which both receptor and binding proteins are found abundantly in the epididymis (Porter et al., 1985; Akmal, 1996), and vitamin D for which receptors are found in epithelial cells of the efferent ducts and caput epididymidis (Schleicher et al., 1989); its active metabolite is synthesized primarily in cauda epididymidis (Kidroni et al., 1983).

The potential for novel, unexpected regulators of epididymal functions has recently surfaced. For example, the mRNAs for MEL1 A and MEL1B melatonin receptors have been identified in principal cells of the corpus epididymidis (Li et al., 1998). Furthermore, high affinity receptors for melatonin are found in this tissue, and receptor binding activity is regulated by testosterone and independent of light-induced changes in circulating melatonin (Shiu et al., 1996).

14.4.3.1 Effects on the Epididymis of Androgen Withdrawal

There is a time dependent dramatic weight loss of the epididymis after androgen withdrawal. This weight loss is in part due to the loss of spermatozoa and fluid from the lumen of the epididymis and in part due to changes in the epithelium of the tissue (Robaire et al, 1977). By using in situ TUNEL apoptosis detection and the presence of DNA ladders by Southern blots, it is clear that apoptotic cell death plays a major role after androgen withdrawal; this process accounts, at least in part, for the loss of epithelial weight (Fan and Robaire, 1998). Apoptotic cell death after orchidectomy first appears in the epithelium of the initial segment of the epididymis and subsequently in more distal segments, i.e. the peak of apoptotic cell death in the initial segment of the epididymidis is reached after two days, whereas that in the cauda epididymidis is not reached until six days post-orchidectomy. Throughout the epididymis, this wave of apoptotic cell death is localized specifically to principal cells. Androgen replacement and efferent duct ligation studies show that while apoptosis in the initial segment is dependent on both androgens and luminal components coming from the testis, in the rest of the epididymis, apoptotic cell death after orchidectomy can be averted by androgens alone.

14.4.3.2 The Active Androgen in the Epididymis

The initial observation that the epididymis is androgen dependent has been extended by several complementary lines of evidence that demonstrate that the main androgen responsible for maintaining epididymal structure and functions is the 5α-reduced metabolite of testosterone, dihydrotestosterone (DHT). The rate limiting enzyme in the pathway leading from testosterone to its 5α-reduced metabolites is 4-ene steroid 5α-reductase (EC 1.3.1.22). Several key findings led to this conclusion. The first is that the active androgen present in epididymal cell nuclei after injection of radiolabelled testosterone is DHT (Tindall, 1972). The second is that epididymal tissue can synthesize 5α-reduced metabolites from testosterone in vitro (Gloyna and Wilson, 1969; Robaire et al., 1977). The third is that results of micropuncture experiments confirm that, beyond the efferent ducts, the predominant androgens in epididymal luminal fluid are 5α-reduced metabolites of testosterone (Turner et al., 1984). The fourth is that 5α-reduced metabolites of testosterone are more potent than testosterone in maintaining epididymal functions *in vitro* (Orgebin-Crist et al., 1976). Finally, there are some data indicating that administration of 5α-reductase inhibitors to adult rats results in diminished epididymal function (Cohen et al., 1981); however, the effects of recently developed selective and potent inhibitors of the two isozymes found in the epididymis have not been extensively studied.

14.4.3.3 Regulation of Epididymal 5α-Reductases

Studies designed to understand the regulation of epididymal 5α-reductase activity have been done at the level of the enzyme activity, proteins, mRNAs and DNA. We will describe key findings at each of these levels of regulation.

Regulation of Epididymal 5α-Reductase Enzyme Activity. 5α-Reductase activity is present in both the nuclear and microsomal fractions of epididymal homogenates (Scheer and Robaire, 1983). Enzyme activity is expressed in a striking positional gradient in the epididymis of the rat and other species (Robaire et al., 1981). The activity associated with the nuclear fraction is highest in the initial segment and declines dramatically as one moves distally along the tissue. The hormonal regulation of epididymal 5α-reductase activity is complex. The enzyme activity found in the nuclear fraction is markedly decreased after bilateral orchidectomy; the simultaneous administration of exogenous testosterone only partially offsets this decrease, even in very high doses designed to mimic the intraluminal androgen concentration (Robaire et al., 1977). Efferent duct ligation and unilateral orchidectomy both result in a dramatic decrease in epididymal 5α-reductase activity, especially in the proximal portion of the tissue (Robaire et al., 1977; Robaire, 1979). Thus, in the initial segment, epididymal 5α-reductase activity is regulated in a paracrine manner by a substance directly entering the epididymis via the efferent ducts, and not via the general circulation (Robaire et al. 1981); this type of regulation has also been noted for immobilin (Ruiz-Bravo, 1988), CRES (Cornwall et al., 1992), and gamma-glutamyl transpeptidase (Palladino and Hinton, 1994) and has been termed as "lumincrine" by Hinton's group (Hinton et al., 1998). Based on the results of a series of endocrine manipulations that altered the nature of the substance(s) entering the epididymis from the testis via the efferent ducts (Robaire et al., 1981; Robaire and Zirkin, 1981; Robaire and Viger, 1995), it can be shown that the paracrine factor regulating nuclear epididymal 5α-reductase activity is of Sertoli cell origin and under the control of testosterone. Several lines of evidence led us to speculate that this paracrine factor is androgen-binding protein (ABP, Robaire and Viger, 1995). In marked contrast to the nuclear activity, microsomal 5α-reductase activity is found throughout the epididymis and is expressed at a lower level (Scheer and Robaire, 1983). The developmental profile of epididymal microsomal 5α-reductase activity reveals that it is coincidental with serum androgens (Scheer and Robaire, 1980); this activity may be regulated by circulating androgens (Robaire et al., 1981).

Regulation of Epididymal mRNAs for 5α-Reductase. Two cDNAs for 5α-reductase, named types 1 and 2 based on their chronological order of identification, are the products of separate genes (Andersson et al., 1989; Normington and Russell,1992). The 5α-reductase type 1 mRNA is regionally distributed along the adult rat epididymis, and its transcript is most abundantly expressed in the initial segment of the epididymis (Viger and Robaire, 1991). The positional gradient in 5α-reductase type 1 mRNA expression in the epididymis is the same as that for nuclear 5α-reductase enzyme activity (Robaire et al., 1981).

The endocrine and developmental regulation of the 5α-reductase type 1 mRNA reveals that: 1) orchidectomy results in a decrease in type 1 mRNA levels in all epididymal segments; 2) high dose exogenous testosterone maintains 5α-reductase type 1 mRNA at control levels in all regions of the epididymis except the initial segment; 3) unilateral orchidectomy (Viger and Robaire, 1991) and efferent duct ligation (Viger and Robaire, 1996) cause a dramatic decrease of type 1 5α-reductase mRNA, selectively in the initial segment of the epididymis (Viger and Robaire, 1991; Viger and Robaire, 1996); and 4) the type 1 transcript is developmentally regulated (Viger and Robaire, 1992). The fact that epididymal 5α-reductase type 1 mRNA concentrations increased during postnatal development is consistent with changes seen in enzyme activity (Scheer and Robaire, 1980). Thus, as for enzyme activity, the primary regulator of 5α-reductase type 1 mRNA expression in the initial segment of the epididymis is a paracrine factor of testicular origin entering the epididymis via the efferent ducts, while 5α-reductase type 1 mRNA levels in the other epididymal regions is controlled by circulating androgens (Viger and Robaire, 1991).

The 5α-reductase type 2 mRNA transcript is 3.6 kb in length, is found in high concentrations in the epididymis, and has a spatial distribution along the adult rat epididymis that is markedly different from the one described for the type 1 mRNA or enzyme activity (Viger and Robaire, 1996). While the 5α-reductase type 1 mRNA expression and enzyme activity are characterized by dramatic increases that occur during postnatal development just before the first appearance of spermatozoa in the epididymis (Scheer and Robaire, 1983; Viger and Robaire, 1992), 5α-reductase type 2 mRNA expression, for all epididymal segments, does not show any significant developmental changes (Viger and Robaire, 1996). Together, these data support our hypothesis that the

5α-reductase type 1 mRNA is the dominant transcript expressed in the epididymis because of the close association between mRNA type 1 levels and enzyme activity, and that the type 2 transcript, though abundant, is not expressing an active enzyme in this tissue.

Regulation of Epididymal 5α-Reductase Protein Expression. In comparing the regulation of epididymal 5α-reductase enzyme activity and the expression of its mRNAs, it is apparent that another potential regulatory site for the enzyme is at the level of the protein itself. It is interesting to note that post-transcriptional regulation has also been proposed for several other epididymal proteins (Cornwall and Hann, 1995b; Brooks et al., 1986; Cyr et al., 1993; Ghyselinck et al., 1990) and other proteins in the male reproductive tract (Kistler et al., 1981; Collins et al, 1988).

In the adult rat, 5α-reductase type 1 protein expression is intensely immunolocalized in discrete lobules of the proximal initial segment of the epididymis (Viger and Robaire, 1994). A sharp decline in staining intensity occurs between the proximal initial segment and its adjacent region, followed by a progressive decrease in intensity. In all epididymal regions, the expression of the 5α-reductase type 1 protein was specific to the epithelial principal cells. This longitudinal and cellular localization is consistent with the reported distribution (Robaire et al., 1977; Viger and Robaire, 1991) and cell-type specificity of 5α-reductase type 1 mRNA (Berman and Russell, 1993) and enzyme activity (Klinefelter and Amann, 1980). No studies on the cellular and subcellular localization of the type 2 isozyme have yet been reported.

Regulation at the Genomic Level. The basic gene structure and chromosomal locations of the human 5α-reductase isozymes have been reported (Jenkins et al.,1991; Labrie et al.,1992, Thigpen et al.,1992). Neither the 5' upstream sequences nor trans or cis-acting factors have been reported for either the human or rat genes. Since the expression of a gene is driven by promoter sequences that are usually located 5' upstream of the initiation codon, we have initiated studies on the isolation and sequencing of the 5' upstream sequences of the two rat genes.

We have obtained a 4.5 kb fragment for type 1 which is localized in the 5' upstream region, while for the type 2 a 2.1 kb fragment, localized in the 5' upstream region, has been obtained. Analysis of both sequences reveals the presence of a large number of presumptive binding sites for described transcription factors that have the potential of playing a role in the activation of the 5α-reductase genes. A preliminary analysis revealed that the most abundant sequences are for NF-1, SP1, OCT-2, AR, GATA, and PIT; other less frequent sequences are for AP-1, AP-2, PEA-3, and CREB. Interestingly, no striking difference was noted between the range of potential factors for the type 1 and type 2 5α-reductase genes.

14.5 The Aging Process: Changes in Structure and Functions of the Epididymis

During aging, the male reproductive tract is characterized by testicular dysfunction resulting in atrophy of the seminiferous epithelium and of the Leydig cells. Although the process of aging has been studied extensively for the testis and the prostate, there is little information on how aging affects the epididymis. The development of the Brown Norway rat as a model to study male reproductive aging (Wright et al., 1993) has been instrumental in gaining insight into the effects of aging on the epididymis. This strain of rat is long lived and is relatively disease free, yet there is a decline in reproductive function when no other system is apparently adversely affected by age, thus allowing for the study of male reproductive aging in the absence of confounding pathologies found in other strains and species, such as pituitary and testicular tumors.

The epididymis is highly dependant on the presence of androgens, particularly 5α-reduced androgens, in order to function. Since steroidogenesis decreases with age in the Brown Norway, as it does in man, one of the first questions to be addressed regarding aging of the epididymis was to examine how specific markers of epididymal function are affected with age. Examination, at the mRNA level, of such markers indicates that epididymal function is not altered in an all or none fashion and furthermore, the effects of age are not restricted to a single morphological region. For example, 5α-reductase type 1 mRNA levels decrease only in the caput-corpus epididymidis, clusterin mRNA levels change only in the cauda epididymidis, and yet androgen receptor mRNA levels and the levels of other androgen dependant messages such as those for protein B/C and D/E, remain unchanged (Viger and Robaire, 1995).

This indicates that while the ability of the epididymis to produce 5α-reduced androgens is compromised in a segment specific manner with age, the ability of the tissue to respond to androgens is not.

Studies on the histology of the epididymis in the Brown Norway rat revealed that, in addition to the emergence of characteristic signs of aging (lipofuscin accumulation and thickening of the basement membrane), the epididymal epithelium undergoes profound cell and segment specific ultrastructural changes with age (Serre and Robaire,1998a). In the initial segment, only the basal cells appear to be primarily affected by age. As early as twelve months of age, these cells begin to emit pseudopodia into the thickening basement membrane. In the caput epididymis, the clear cells are the most modified with age. By 18 months, these cells are swollen and had lost some of their characteristic subcellular compartmentalization. In the corpus, advancing age primarily effects principal cells, which by 18 months show remarkable increases in the size and number of lysosomes. In the proximal cauda epididymidis, aging results in the most dramatic change in appearance; clear cells are enlarged and filled with dense lysosomes and some principal cells contained large vacuoles. The molecular mechanisms behind these changes remain unclear, however, many hypotheses have been suggested, including; dysregulated intracellular trafficking, decreased protein degradation, and oxidative stress. Oxidative stress is widely thought to play a key, or potentially causative role in the pathogenesis of aging (Ashok and Ali, 1999). Immunohistochemical localization studies of glutathione-S-transferases (Ya, Yc, Yb1, Yb2, Yo and Yf) in the aging Brown Norway rat reveal that aging alters their expression in a region specific manner (Mueller et al., 1998). More specifically, the expression of GSTs is affected only in principal cells of the proximal cauda epididymis. By 24 months, normal principal cells in this region loose their ability to express Ya but maintained expression of Yc, Yo, Yb1 and Yb2. In contrast, those principal cells, which had become vacuolated, maintain the ability to express Ya but no longer expressed Yc, Yo, Yb1 or Yb2.

Aging in the epididymis is also accompanied by an activation of the immune system as evidenced by increased numbers of intra-epithelial halo cells and the emergence of two distinct populations of halo cells (Serre and Robaire, 1999). Since the blood-epididymis barrier should protect from immunological attack, these results indicated that the structure and function of the blood epididymis barrier may be compromised with age. Immunolocalization studies of proteins involved in maintaining the integrity of tight junctions, occludin, ZO-1, and E-cadherin, indicate that there are segment specific changes in the expression and distribution of these proteins with age (Levy and Robaire, 1999). The effects were particularly striking in the corpus epididymidis. Using an electron opaque tracer, lanthanum nitrate, the permeability of the barrier was also demonstrated to be compromised with age. The greatest increase in lanthanum permeability occurred in the corpus epididymidis, corresponding with the changes observed for the junctional proteins. Thus it appears that both the structure and function of the barrier are affected by aging.

Using antibodies specific to various immune cell populations (helper T lymphocytes, cytotoxic T lymphocytes, B lymphocytes and monocyte-macrophages), it can be shown that the age of the animal, the segment and the luminal content all affect the complement of immune cells present in the epithelium of the aging Brown Norway epididymis (Serre and Robaire, 1999). By 18 months, the number of monocytes/macrophages is dramatically increased in the initial segment, caput, and corpus epididymidis. Furthermore, this recruitment is enhanced in aged rats where there are few spermatozoa in the lumen. In contrast, in the cauda epididymidis, monocyte/macrophage and T lymphocyte populations do not change significantly with age. The reason for the diminished effect of age on immune cell recruitment in the cauda epididymidis is unclear.

A functional endpoint of epididymal function is the ability of spermatozoa to acquire normal motility characteristics. When the acquisition of sperm motility along the epididymis of Brown Norway rats is compared, using computer-assisted sperm analysis, between young and old animals, several kinematic parameters of sperm motility in the cauda epididymidis were decreased. However, no changes were noted for sperm found in the caput epididymidis. This altered acquisition of sperm motility probably results from major epididymal dysfunctions. In addition, it is clear that when young females are mated to older males, there are higher incidences of pre-implantation loss, lower fetal weights, and higher post-natal death rates when compared to matings with young males (Serre and Robaire, 1998b). Whether these effects are due to altered epididymal maturation of spermatozoa or to changes taking place in their development while in the testis, is yet to be resolved.

Acknowledgements. Studies from our laboratory presented above were funded by grants from the MRC and NIH (NIA 08321). PS is the recipient of a France-Quebec Exchange Fellowship.

References

Abou-Haila A, Fain-Maurel MA (1985) Postnatal differentiation of the enzymatic activity of the mouse epididymis. Int J Androl 8:441–458

Akmal KM, Dufour JM, Kim KH (1996) Region-specific localization of retinoic acid receptor-alpha expression in the rat epididymis. Biol Reprod 54:1111–1119

Andersson S, Bishop RW, Russell DW (1989) Expression cloning and regulation of steroid 5α-reductase, an enzyme essential for male sexual differentiation. J Biol Chem 264:16249–16255

Andonian S, Badran H, Hermo L (1999a) Immunolocalization of acquaporins (AQPs) 6 and 9 in the adult rat vas deferens (VD). Cell Biol meeting 1610

Andonian S, Hermo L (1999b) Cell- and region-specific localization of lysosomal and secretory proteins and endocytic receptors in epithelial cells of the cauda epididymidis and vas deferens of the adult rat. J Androl 20:415–429

Ashok BT, Ali R (1999) The aging paradox: free radical theory of aging. Exp Gerontol 34:293–303

Ayer-LeLievre C, Olson L, Ebendal T, Hallbook F, Persson H (1988) Nerve growth factor mRNA and protein in the testis and epididymis of mouse and rat. Proc Natl Acad Sci USA 85:2628–2632

Baker CS, Magargee SF, Hammerstedt RH (1993) Cholesterol transfer proteins from ram caudal epididymal and seminal plasma. Biol Reprod Suppl 48:86

Barbieri MA, Veisaga ML, Paolicchi F, Fornes MW, Sosa MA, Mayorga LS, Bustos-Obregon E, Bertini F (1996) Affinity sites for beta-glucuronidase on the surface of human spermatozoa. Andrologia 28:327–333

Bedford JM (1967) Effects of duct ligation on the fertilizing ability of spermatozoa from different regions of the rabbit epididymis. J Exp Zool 166:271–282

Belmonte S, Maturano M, Bertini MF, Pusiol E, Sartor T, Sosa MA (1998) Changes in the content of rat epididymal fluid induced by prolonged treatment with tamoxifen. Andrologia 30:345–350

Benoit J (1926) Recherches anatomiques, cytologiques et histophysiologiques sur les voies excretrices du testicule chez les mammifères. Arch Anat Histol Embryol (Strasb) 5:175–412

Berman DM, Russell DW (1993) Cell-type-specific expression of rat steroid 5α-reductase isozymes. Proc Natl Acad Sci USA 90:9359–9363

Bérubé B, Sullivan R (1994) Inhibition of in vivo fertilization by active immunization of male hamsters against a 26-kDa sperm glycoprotein. Biol Reprod 51:1255–1263

Blanchard Y, Robaire B (1997) Le mode d'action des androgènes et la 5α-réductase. Médecine/Sciences 13:467–473

Brooks DE (1981) Secretion of proteins and glycoproteins by the rat epididymis: regional differences, androgen-dependence, and effects of protease inhibitors, procaine, and tunycamycin. Biol Reprod 25:1099–1117

Brooks DE (1983a) Influence of incubation conditions, tunicamycin and castration on incorporation of [3H]mannose and [3H]fucose into rat epididymal glycoproteins in vitro. J Reprod Fertil 67:97–105

Brooks DE (1983b) Effect of androgens on protein synthesis and secretion in various regions of the rat epididymis, as analysed by two-dimensional gel electrophoresis. Mol Cell Endocrinol 29:255–270.

Brooks DE, Means AR, Wright EJ, Singh SP, Tiver KK (1986) Molecular cloning of the cDNA for androgen-dependent sperm-coating glycoproteins secreted by the rat epididymis. Eur J Biochem 161:13–18

Byskov AG, Høyer PE (1988) Embryology of mammalian gonads and ducts. In: Knobil E, Neill J (eds) The Physiology of Reproduction. Raven Press, New York, pp 265–302

Cameo MS, Blaquier JA (1976) Androgen-controlled specific proteins in rat epididymis. J Endocrinol 69:47–55

Chaturapanich G, Jones RC, Culow J (1992) Protein synthesis and secretion by the epididymis of Tammar Wallaby, Macropus eugenii. Reprod Fert Dev 4:533–545

Chen MY, Carpenter D, Zhao GQ (1999) Expression of bone morphogenetic protein 7 in murine epididymis is developmentally regulated. Biol Reprod 60:1503–1508

Cheng CY, Mathur PP, Grima J (1988) Structural analysis of clusterin and its subunits in ram rete testis fluid. Biochemistry 27:4079–4088

Clermont Y, Flannery J (1970) Mitotic activity in the epithelium of the epididymis in young and old adult rats. Biol Reprod 3:283–292

Cohen J, Ooms MP, Vreeburg JTM (1981) Reduction of fertilizing capacity of epididymal spermatozoa by 5α-steroid reductase inhibitors. Experientia 37:1031–1032

Collins S, Quarmby V, French F, Lefkowitz RJ, Caron MG (1988) Regulation of the β_2-adrenergic receptor in the rat ventral prostate by testosterone. FEBS Lett 233:173–176

Cooper TG (1990) In defense of a function for the human epididymis. Fert Steril 54:965–975

Cornwall GA, Hann SR (1995a) Transient appearance of CRES protein during spermatogenesis and caput epididymal sperm maturation. Mol Reprod Dev 41:37–46

Cornwall GA, Hann SR (1995b) Specialized gene expression in the epididymis. J Androl 16:379–383

Cornwall GA, Hsia N (1997) ADAM7, a member of the ADAM (a disintegrin and metalloprotease) gene family is specifically expressed in the mouse anterior pituitary and epididymis. Endocrinology 138:4262–4272

Cornwall GA, Orgebin-Crist MC, Hann SR (1992) The CRES gene: a unique testis-regulated gene related to the cystatin family is highly restricted in its expression to the proximal region of the mouse epididymis. Mol Endocrinol 6:1653–1664

Cornwall GA, Tulsiani DR, Orgebin-Crist MC (1991) Inhibition of the mouse sperm surface alpha-D-mannosidase inhibits sperm-egg binding in vitro. Biol Reprod 44:913–921

Couse JF, Lindzey J, Grandien K, Gustafsson JA, Korach KS (1997) Tissue distribution and quantitative analysis of estrogen receptor-alpha (ERalpha) and estrogen receptor-beta (ERbeta) messenger ribonucleic acid in the wild-type and ERalpha-knockout mouse. Endocrinology 138:4613–1621

Cyr DG, Hermo L, Blaschuk OW, Robaire B (1992) Distribution and regulation of epithelial cadherin messenger ribonucleic acid and immunocytochemical localization of epithelial cadherin in the rat epididymis. Endocrinology 130:353–363

Cyr DG, Hermo L, Egenberger N, Mertineit C, Trasler JM, Laird DW (1999) Cellular immunolocalization of occludin during embryonic and postnatal development of the mouse testis and epididymis. Endocrinology 140:3815–3825

Cyr DG, Hermo L, Robaire B (1993) Developmental changes in epithelial cadherin messenger ribonucleic acid and immunocytochemical localization of epithelial cadherin during postnatal epididymal development in the rat. Endocrinology 132:1115–1124

Cyr DG, Robaire B (1992) Regulation of sulfated glycoprotein-2 (clusterin) messenger ribonucleic acid in the rat epididymis. Endocrinology 130:2160–2166

Dacheux JL, Dacheux F, Paquignon M (1989) Changes in sperm surface membrane and luminal protein fluid content during epididymal transit in the boar. Biol Reprod 40:635–651

Desai KV, Flanders KC, Kondaiah P (1998) Expression of transforming growth factor-beta isoforms in the rat male accessory sex organs and epididymis. Cell Tissue Res 294:271–277

Del-Rio AG (1979) Macromolecular secretion into various segments of the guinea pig epididymis. Arch Androl 3:231–237

Drevet JR, Lareyre JJ, Schwaab V, Vernet P, Dufaure JP (1998) The PEA3 protein of the Ets oncogene family is a putative transcriptional modulator of the mouse epididymis-specific glutathione peroxidase gene GPX5. Mol Reprod Dev 49:131–140

Fan X, Robaire B (1998) Orchidectomy induces a wave of apoptotic cell death in the epididymis. Endocrinology 139:2128–2136

Fisher JS, Turner KJ, Fraser HM, Saunders PT, Brown D, Sharpe RM (1998) Immunoexpression of aquaporin-1 in the efferent ducts of the rat and marmoset monkey during development, its modulation by estrogens, and its possible role in fluid resorption. Endocrinology 139:3935–3945

Flickinger CJ, Bush LA, Howards SS, Herr JC (1997) Distribution of leucocytes in the epithelium and interstitium of four regions of the Lewis rat epididymis. Anat Rec 248:380–390

Garrett JE, Garrett SH, Douglass J (1990) A spermatozoa-associated factor regulates proenkephalin gene expression in the rat epididymis. Mol Endocrinol 4:108–118

Ghyselinck NB, Jimenez C, Lefrancois AM, Dufaure JP (1990) Molecular cloning of a cDNA for androgen-regulated proteins secreted by the mouse epididymis. J Mol Endocrinol 4:5–12

Gloyna RE, Wilson JD (1969) A comparative study of the conversion of testosterone to 17β-hydroxy-5α-androstan-3-one (dihydrotestosterone) by rat prostate and epididymis. J Clin Endocrinol Metab 29:970–977

Hamilton DW (1980) UDP-galactose:N-acetylglucosamine galactosyltransferase in fluids from rat rete testis and epididymis. Biol Reprod 23:377–385

Hall JC, Perez FM, Kochins JG, Pettersen CA, Li Y, Tubbs CE, LaMarche MD (1996) Quantification and localization of N-acetyl-beta-D-hexosaminidase in the adult rat testis and epididymis. Biol Reprod 54:914–929

Hermo L, Barin K, Robaire B (1992a) Structural differentiation of the epithelial cells of the testicular excurrent duct system of rats during postnatal development. Anat Rec 233:205–228

Hermo L, Oko R, Robaire B (1992b) Epithelial cells of the epididymis show regional variations with respect to the secretion of endocytosis of immobilin as revealed by light and electron microscope immunocytochemistry. Anat Rec 232:202–220

Hermo L, Lustig M, Lefrancois S, Argraves WS, Morales CR (1999) Expression and regulation of LRP-2/megalin in epithelial cells lining the efferent ducts and epididymis during postnatal development. Mol Reprod Dev 53:282–293

Hermo L, Wright J, Oko R, Morales CR (1991) Role of epithelial cells of the male excurrent duct system of the rat in the endocytosis or secretion of sulfated glycoprotein-2 (clusterin). Biol Reprod 44:1113–1131

Hess RA, Gist DH, Bunick D, Lubahn DB, Farrell A, Bahr J, Cooke PS, Greene GL (1997) Estrogen receptor (alpha and beta) expression in the excurrent ducts of the adult male rat reproductive tract. J Androl 18:602–611

Hinton BT, Lan ZJ, Rudolph DB, Labus JC, Lye RJ (1998) Testicular regulation of epididymal gene expression. J Reprod Fertil Suppl 53:47–57

Hinton BT, Palladino MA, Rudolph D, Labus JC (1995) The epididymis as protector of maturing spermatozoa. Reprod Fertil Dev 7:731–745

Holland MK, Orgebin-Crist MC (1988) Characterization and hormonal regulation of protein synthesis by the murine epididymis. Biol Reprod 38:487–496

Igdoura SA, Herscovics A, Lal A, Moremen KW, Morales CR, Hermo L (1999) Alpha-mannosidases involved in N-glycan processing show cell specificity and distinct subcompartmentalization within the Golgi apparatus of cells in the testis and epididymis. Eur J Cell Biol 78:441–452

Igdoura SA, Morales CR, Hermo L (1995) Differential expression of cathepsins B and D in testis and epididymis of adult rats. J Histochem Cytochem 43:545–557

Janulis L, Bahr JM, Hess RA, Janssen S, Osawa Y, Bunick D (1998) Rat testicular germ cells and epididymal sperm contain active P450 aromatase. J Androl 19:65–71

Jauhiainen A, Vanha-Perttula T (1985) Acid and neutral alpha-glucosidase in the reproductive organs and seminal plasma of the bull. J Reprod Fertil 74:669–680

Jeannotte L, Burbach JP, Drouin J (1987) Unusual proopiomelanocortin ribonucleic acids in extrapituitary tissues: intronless transcripts in testes and long poly(A) tails in hypothalamus. Mol Endocrinol 1:749–757

Jenkins E, Hsieh C-L, Milatovich A, Normington K, Berman DM, Francke U, Russell DW (1991) Characterization and chromosomal mapping of a human steroid 5α-reductase gene and pseudogene and mapping of the mouse homologue. Genomics 11:1102–1112

Jin YZ, Bannai S, Dacheux F, Dacheux JL, Okamura N (1997) Direct evidence for the secretion of lactoferrin and its binding to sperm in the porcine epididymis. Mol Reprod Dev 47:490–496

Kaunisto K, Fleming RE, Kneer J, Sly WS, Rajaniemi H (1999) Regional expression and androgen regulation of carbonic anhydrase IV and II in the adult rat epididymis. Biol Reprod 61:1521–1526

Kidroni G, Har-Nir R, Menezel J, Frutkoff IW, Palti Z, Ron M (1983) Vitamin D3 metabolites in rat epididymis: high 24,25-dihydroxy vitamin D3 levels in the cauda region. Biochem Biophys Res Commun 113:982–989

Kirchhoff C (1999) Gene expression in the epididymis. Int Rev Cytol 188:133–202

Kirchhoff C, Habben I, Ivell R, Krull N (1991) A major human epididymis-specific cDNA encodes a protein with sequence homology to extracellular proteinase inhibitors. Biol Reprod 45:350–357

Kirchhoff C, Krull N, Pera I, Ivell R (1993) A major mRNA of the human epididymal principal cells, HE5, encodes the leucocyte differentiation CDw52 antigen peptide backbone. Mol Reprod Dev 34:8–15

Kirchhoff C, Osterhoff C, Young L (1996) Molecular cloning and characterization of HE1, a major secretory protein of the human epididymis. Biol Reprod 54:847–856

Kistler MK, Ostrowski MC, Kistler WS (1981) Developmental regulation of secretory protein synthesis in the rat seminal vesicle. Proc Natl Acad Sci USA 78:737–741

Klinefelter GR, Amann RP (1980) Metabolism of testosterone by principal cells and basal cells isolated from the rat epididymal epithelium. Biol Reprod 22:1149–1154

Labrie F, Sugimoto Y, Luu-The V, Simard J, Lachance Y, Bachvarov D, Leblanc G, Durocher F, Paquet N (1992) Structure of human type II 5α-reductase gene. Endocrinology 131:1571–1573

Lan ZJ, Labus JC, Hinton BT (1998) Regulation of gamma-glutamyl transpeptidase catalytic activity and protein level in the initial segment of the rat epididymis by testicular factors: role of basic fibroblast growth factor. Biol Reprod 58:197–206

Lan ZJ, Lye RJ, Holic N, Labus JC, Hinton BT (1999) Involvement of polyomavirus enhancer activator 3 in the regulation of expression of gamma-glutamyl transpeptidase messenger ribonucleic acid-IV in the rat epididymis. Biol Reprod 60:664–673

Lareyre JJ, Thomas TZ, Zheng WL, Kasper S, Ong DE, Orgebin-Crist MC, Matusik RJ (1999) A 5-kilobase pair promoter fragment of the murine epididymal retinoic acid-binding protein gene drives the tissue-specific, cell-specific, and androgen-regulated expression of a foreign gene in the epididymis of transgenic mice. J Biol Chem 274:8282–8290

Lea OA, French FS (1981) Characterization of an acidic glycoprotein secreted by principal cells of the rat epididymis. Biochim Biophys Acta 668:370–376

Legaré C, Gaudreault C, St-Jacques S, Sullivan R (1999) P34H sperm protein is preferentially expressed by the human corpus epididymidis. Endocrinology 140:3318–3327

Levy S, Robaire B (1999) Segment-specific changes in the expression of junctional proteins and the permeability of the blood-epididymis barrier with age. Biol Reprod 60:1392–1401

Li L, Xu JN, Wong YH, Wong JT, Pang SF, Shiu SY (1998) Molecular and cellular analyses of melatonin receptor-mediated cAMP signaling in rat corpus epididymis. J Pineal Res 25:219–228

Lindsey S, Wilkinson MF (1996) Homeobox genes and male reproductive development. J Assist Repro Genetics 13:182–192

Mattmueller DR, Hinton BT (1991) In vivo secretion and association of clusterin (SGP-2) in luminal fluid with spermatozoa in the rat testis and epididymis. Mol Reprod Dev 30:62–69

Mayorga LS, Bertini F (1985) The origin of some acid hydrolases of the fluid of the rat cauda epididymidis. J Androl 6:243–245

Miranda PV, Brandelli A, Tezon JG (1995) Characterization of beta-N-acetylglucosaminidase from human epididymis. Int J Androl 18:263–270

Mueller A, Hermo L, Robaire B (1998) The effects of aging on the expression of glutathione S-transferases in the testis and epididymis of the Brown Norway rat. J Androl 19:450–465

Normington K, Russell DW (1992) Tissue distribution and kinetic characteristics of rat steroid 5α-reductase isozymes. J Biol Chem 267:19548–19554

Oefelein M, Grapey D, Schaeffer T, Chin-Chance C, Bushman W (1996) Pax-2: a developmental gene constitutively expressed in the mouse epididymis and ductus deferens. J Urol 156:1204–1207

Okamura N, Dacheux F, Venien A, Onoe S, Huet JC, Dacheux JL (1992) Localization of a maturation-dependent epididymal sperm surface antigen recognized by a monoclonal antibody raised against a 135-kilodalton protein in porcine epididymal fluid. Biol Reprod 47:1040–1052

Okamura N, Iwaki Y, Hiramoto S, Tamba M, Bannai S, Sugita Y, Syntin P, Dacheux F, Dacheux JL (1997) Molecular cloning and characterization of the epididymis-specific glutathione peroxidase-like protein secreted in the porcine epididymal fluid. Biochim Biophys Acta 1336:99–109

Okamura N, Kiuchi S, Tamba M, Kashima T, Hiramoto S, Baba T, Dacheux F, Dacheux JL, Sugita Y, Jin YZ (1999) A porcine homolog of the major secretory protein of human epididymis, HE1, specifically binds cholesterol. Biochim Biophys Acta 1438:377–387

Okamura N, Tamba M, Liao HJ, Onoe S, Sugita Y, Dacheux F, Dacheux JL (1995b) Cloning of complementary DNA encoding a 135-kilodalton protein secreted from porcine corpus epididymis and its identification as an epididymis-specific alpha-mannosidase. Mol Reprod Dev 42:141–148

Okamura N, Tamba M, Uchiyama Y, Sugita Y, Dacheux F, Syntin P, Dacheux JL (1995a) Direct evidence for the elevated synthesis and secretion of procathepsin L in the distal caput epididymis of boar. Biochim Biophys Acta 1245:221–226

Olson GE, Hinton BT (1985) Regional differences in luminal fluid polypeptides of the rat testis and epididymis revealed by two-dimensional gel electrophoresis. J Androl 6:20–34

Ong DE, Chytil F (1988) Presence of novel retinoic acid-binding proteins in the lumen of rat epididymis. Arch Biochem Biophys 267:474–478

Orgebin-Crist M-C (1967) Sperm maturation in rabbit epididymis. Nature 216:816–818

Orgebin-Crist MC, Djiane J (1979) Properties of a prolactin receptor from the rabbit epididymis. Biol Reprod 21:135–139

Orgebin-Crist M-C, Jahad N, Hoffman LH (1976) The effects of testosterone, 5α-dihydrotestosterone, 3α-androstandiol, and 3β-androstandiol on the maturation of rabbit epididymal spermatozoa in organ culture. Cell Tissue Res 167:515–525

Osterhoff C, Kirchhoff C, Krull N, Ivell R (1994) Molecular cloning and characterization of a novel human sperm antigen (HE2) specifically expressed in the proximal epididymis. Biol Reprod 50:516–525

Palladino MA, Hinton BT (1994) Expression of multiple gamma-glutamyl transpeptidase messenger ribonucleic acid transcripts in the adult rat epididymis is differentially regulated by androgens and testicular factors in a region-specific manner. Endocrinology 135:1146–1156

Papp S, Robaire B, Hermo L (1995) Immunocytochemical localization Ya, Yc, Yb_1, and Yb_2 subunits of glutathione S-transferases in the testis and epididymis of adult rats. Micro Res Tech 30:1–23

Pera I, Derr P, Yeung CH, Cooper TG, Kirchhoff C (1997) Regionalized expression of CD52 in rat epididymis is related to mRNA poly(A) tail length. Mol Reprod Dev 48:433–441

Perry AC, Jones R, Barker PJ, Hall L (1992) A mammalian epididymal protein with remarkable sequence similarity to snake haemorrhagic peptides. Biochem J 286:671–675

Perry AC, Jones R, Hall L (1993a) Isolation and characterization of a rat cDNA clone encoding a secreted superoxide dismutase reveals the epididymis to be a major site of its expression. Biochem J 293:21–25

Perry AC, Jones R, Hall L (1993b) Sequence analysis of monkey acrosin-trypsin inhibitor transcripts and their abundant expression in the epididymis. Biochim Biophys Acta 1172:159–160

Podlasek CA, Seo RM, Clemens JQ, Ma L, Maas RL, Bushman W (1999) Hoxa-10 deficient male mice exhibit abnormal development of the accessory sex organs. Dev Dynamics 214:1–12

Porter SB, Ong DE, Chytil F, Orgebin-Crist MC (1985) Localization of cellular retinol-binding protein and cellular retinoic acid-binding protein in the rat testis and epididymis. J Androl 6:197–212

Rankin TL, Tsuruta KJ, Holland MK, Griswold MD, Orgebin-Crist MC (1992) Isolation, immunolocalization, and sperm-association of three proteins of 18, 25, and 29 kilodaltons secreted by the mouse epididymis. Biol Reprod 46:747–766

Ravet V, Depeiges A, Morel F, Dufaure JP (1991) Synthesis and post-translational modifications of an epididymal androgen dependent protein family. Gen Comp Endocrinol 84:104–114

Robaire B (1979) Effects of unilateral orchidectomy on rat epididymal Δ^4–5α-reductase and 3α-hydroxysteroid dehydrogenase. Can J Physiol Pharmol 57:998–100

Robaire B, Ewing LL, Zirkin BR, Irby DC (1977) Steroid Δ^4–5α-reductase and 3α-hydroxysteroid dehydrogenase in the rat epididymis. Endocrinology 101:1379–1390

Robaire B, Hermo L (1988) Efferent ducts, epididymis, and vas deferens: structure, functions, and their regulation. In: Knobil E, Neill J (eds) The Physiology of Reproduction. Raven Press, New York, pp 999–1080

Robaire B, Scheer H, Hachey C (1981) Regulation of epididymal steroid metabolizing enzymes. In: Jagiello G, Vogel HJ (eds) Bioregulators of Reproduction. Academic Press, New York, pp 487–498

Robaire B, Viger R (1995) Regulation of epididymal epithelial cell functions. Biol Reprod 52:226–236

Robaire B, Zirkin BR (1981) Hypophysectomy and simultaneous testosterone replacement: effects on male rat reproductive tract and epididymal Δ^4–5α-reductase and 3α-hydroxysteroid dehydrogenase. Endocrinology 109:1225–1233

Rockwerger I, Cohen D, Cuasnicu PS (1992) Mammalian sperm-egg fusiom: the rat egg has complementary sites for a sperm protein that mediate gamete fusion. Dev Biol 153:83–90

Ruiz-Bravo N (1988) Tissue and cell specificity of immobilin biosynthesis. Biol Reprod 39:901–911

Scheer H, Robaire B (1980) Steroid Δ^4–5α-reductase and 3α-hydroxysteroid dehydrogenase in the rat epididymis during postnatal development. Endocrinology 107:948–953

Scheer H, Robaire B (1983) Subcellular distribution of steroid Δ^4–5α-reductase and 3α-hydroxysteroid dehydrogenase in the rat epididymis during sexual maturation. Biol Reprod 29:1–10

Schleicher G, Privette TH, Stumpf WE (1989) Distribution of soltriol [1,25(OH)2-vitamin D3] binding sites in male sex organs of the mouse: an autoradiographic study. J Histochem Cytochem 37:1083–1096

Schwaab V, Faure J, Dufaure JP, Drevet JR (1998) GPx3: the plasma-type glutathione peroxidase is expressed under androgenic control in the mouse epididymis and vas deferens. Mol Reprod Dev 51:362–372

Serre V, Robaire B (1998a) Segment specific morphological changes in the aging Brown Norway rat epididymis. Biol Reprod 58:497–513
Serre V, Robaire B (1998b) Paternal age affects fertility and progeny outcome in the Brown Norway rat. Fertil Steril 70:625–31
Serre V, Robaire, B., (1999) The distribution of immune cells in the epithelium of the epididymis of the aging brown norway rat is segment-specific and related to the luminal content. Biol Reprod 61:705–714
Shiu SY, Chow PH, Yu ZH, Tang F, Pang SF (1996) Autoradiographic distribution and physiological regulation of 2-[125I]iodomelatonin binding in rat epididymis. Life Sci 59:1165–1174
Silber SJ (1989) Results of microsurgical vasoepididymostomy: role of epididymis in sperm maturation. Hum Reprod 4:298–303
Sonnenberg-Riethmacher E, Walter B, Riethmacher D, Godecke S, Birchmeier C (1996) The c-ros tyrosine kinase receptor controls regionalization and differentiation of epithelial cells in the epididymis. Genes Dev 10:1184–1193
Sun EL, Flickinger CJ (1979) Development of cell types and of regional differences in the postnatal rat epididymis. Am J Anat 154:27–55
Sutton KA, Maiti S, Tribley WA, Lindsey JS, Meistrich ML, Bucana CD, Sanborn BM, Joseph DR, Griswold MD, Cornwall GA, Wilkinson MF (1998) Androgen regulation of the Pem homeodomain gene in mice and rat Sertoli and epididymal cells. J Androl 19:21–30
Sylvester SR, Morales C, Oko R, Griswold MD (1991) Localization of sulfated glycoprotein-2 (clusterin) on spermatozoa and in the reproductive tract of the male rat. Biol Reprod 45:195–207
Sylvester SR, Skinner MK, Griswold MD (1984) A sulfated glycoprotein synthesized by Sertoli cells and by epididymal cells is a component of the sperm membrane. Biol Reprod 31:1087–1101
Syntin P, Dacheux JL, Dacheux F (1999) Postnatal development and regulation of proteins secreted in the boar epididymis. Biol Reprod 61:1622–1635
Syntin P, Dacheux F, Druart X, Gatti JL, Okamura N, Dacheux JL (1996) Characterization and identification of proteins secreted in the various regions of the adult boar epididymis. Biol Reprod 55:956–974
Takada M, Yonezawa N, Yoshizawa M, Noguchi S, Hatanaka Y, Nagai T, Kikuchi K, Aoki H, Nakano M (1994) pH-sensitive dissociation and association of beta-N-acetylhexosaminidase from boar sperm acrosome. Biol Reprod 50:860–868
Thigpen AE, Davis DL, Milatovich A, Mendonca BB, Imperato-McGinley J, Griffin JE, Francke U, Wilson JD, Russell DW (1992) Molecular genetics of steroid 5α-reductase 2 deficiency. J Clin Invest 90:799–809
Tindall DJ, French FS, Nayfeh SN (1972) Androgen uptake and binding in rat epididymal nuclei, *in vivo*. Biochem Biophys Res Commun 49:1391–1397

Tomomasa H, Waguri S, Umeda T, Koiso K, Kominami E, Uchiyama Y (1994) Lysosomal cysteine proteinases in rat epididymis. J Histochem Cytochem 42:417–425

Tulsiani DR, NagDas SK, Skudlarek MD, Orgebin-Crist MC (1995b) Rat sperm plasma membrane mannosidase: localization and evidence for proteolytic processing during epididymal maturation. Dev Biol 167:584–595

Tulsiani DR, Orgebin-Crist MC, Skudlarek MD (1998) Role of luminal fluid glycosyltransferases and glycosidases in the modification of rat sperm plasma membrane glycoproteins during epididymal maturation. J Reprod Fertil Suppl 53:85–97

Tulsiani DR, Skudlarek MD, Araki Y, Orgebin-Crist MC (1995a) Purification and characterization of two forms of beta-D-galactosidase from rat epididymal luminal fluid: evidence for their role in the modification of sperm plasma membrane glycoprotein(s). Biochem J 305:41–50

Tulsiani DR, Skudlarek MD, Holland MK, Orgebin-Crist MC (1993b) Glycosylation of rat sperm plasma membrane during epididymal maturation. Biol Reprod 48:417–428

Tulsiani DR, Skudlarek MD, Nagdas SK, Orgebin-Crist MC (1993a) Purification and characterization of rat epididymal-fluid alpha-D-mannosidase: similarities to sperm plasma-membrane alpha-D-mannosidase. Biochem J 290:427–436

Turner TT, Jones CE, Howards SS, Ewing LL, Zegeye B, Gunsalus GL (1984) On the androgen microenvironment of maturing spermatozoa. Endocrinology 115:1925–1932

Turner TT, Miller DW, Avery EA (1995) Protein synthesis and secretion by the rat caput epididymis in vivo: influence of the luminal microenvironment. Biol Reprod 52:1012–1019

Ueda H, Hirano T, Fujimoto S (1990) Changes in proteins secretion patterns during the development of the rat epididymis. Zool Sci 55:119–125

Veri JP, Hermo L, Robaire B (1993) Immunocytochemical localization of the Y_f subunit of glutathione S-transferase P shows regional variation in the staining of epithelial cells of the testis, efferent ducts and epididymis of the male rat. J Androl 14:23–44

Vierula ME, Rankin TL, Orgenin-Crist M-C (1995) Electron microscopic immunolocalization of the 18 and 29 kilodalton secretory proteins in the mouse epididymis: evidence for differential uptake by clear cells. Micro Res Tech 30:24–36

Viger RS, Robaire B (1991) Differential regulation of steady state 4-ene steroid 5α-reductase messenger ribonucleic acid levels along the rat epididymis. Endocrinology 128:2407–2414

Viger RS, Robaire B (1992) Expression of 4-ene steroid 5α-reductase messenger ribonucleic acid in the rat epididymis during postnatal development. Endocrinology 131:1534–1540

Viger RS, Robaire B (1994) Immunocytochemical localization of 4-ene steroid 5α-reductase type 1 along the rat epididymis during postnatal development. Endocrinology 134:2298–2306

Viger RS, Robaire B (1995) Gene expression in the aging Brown Norway rat epididymis. J Androl 16:108–117

Viger RS, Robaire B (1996) The mRNAs for the steroid 5α-reductase isozymes, type 1 and type 2, are differentially regulated in the rat epididymis. J Androl 17:27–34

Wichmann L, Vaalasti A, Vaalasti T, Tuohimaa P (1989) Localization of lactoferrin in the male reproductive tract. Int J Androl 12:179–186

Winer MA, Wolgemuth DJ (1995) The segment-specific pattern of A-raf expression in the mouse epididymis is regulated by testicular factors. Endocrinology 136:2561–2572

Wong PY, Chan HC, Leung PS, Chung YW, Wong YL, Lee WM, Ng V, Dun NJ (1999) Regulation of anion secretion by cyclo-oxygenase and prostanoids in cultured epididymal epithelia from the rat. J Physiol 514:809–820

Wright WW, Fiore C, Zirkin BR. (1993) The effect of aging on the seminiferous epithelium of the Brown Norway rat. J Androl 14:110–117

Xu W, Ensrud KM, Hamilton DW (1997) The 26 kD protein recognized on rat cauda epididymal sperm by monoclonal antibody 4E9 has internal peptide sequence that is identical to the secreted form of epididymal protein E. Mol Reprod Dev 46:377–382

Young WC (1929a) A study of the function of the epididymis. I. Is the attainment of full spermatozoon maturity attributable to some specific action of the epididymal secretion? J Morphol Physiol 47:479–495

Young WC (1929b) A study of the function of the epididymis. II. The importance of an aging process in sperm for the length of the period during which fertilizing capacity is retained by sperm isolated in the epididymis of the guinea-pig. J Morphol Physiol 48:475–491

Young WC (1931) A study of the function of the epididymis. III. Functional changes undergone by spermatozoa during their passage through the epididymis and vas deferens in the guinea pig. J Exp Biol 8:151–162

Yu LC, Chen (1993) The developmental profile of lactoferrin in mouse epididymis. Biochem J 296:107–111

Zhao GQ, Liaw L, Hogan BL (1998) Bone morphogenetic protein 8 A plays a role in the maintenance of spermatogenesis and the integrity of the epididymis. Development 125:1103–1112

Subject Index

activator of CREM in testis 108, 117
activin 98
activin A 95
activin C 99
affinity chromatography 48
aging 248
Alzheimer's disease 44
androgen receptors 91, 92, 165, 174, 180, 187
androgens 165, 173, 178, 179, 187, 189, 243
anti-proliferative properties 47
apoptosis 44, 47, 99, 141
apoptotic systems 144
ARA70 167

basic fibroblast growth factor 95
Bax 99
Bcl-2 99, 132, 135, 146
Bcl-w 99
Bcl-X_L 135

c-kit 97
c-kit receptor 80
Caenorhabditis elegans 41
caspase-independent cell death 133
caspases 148
CBP 108, 110, 112, 114, 117, 121, 123
CBP/p300 170
CDK inhibitors 45
– 9-hydroxyellipticine 45
– γ-butyrolactone 45
– cellular effects 47
– CGP60474 45
– CVT-313 45
– dimethylaminopurine 45
– flavopiridol 45
– hymenialdisine 45
– indirubin-3'-monoxime 45
– isopentenyladenine 45
– olomoucine 45
– paullones 45
– potential applications 49
– purvalanol 45
– roscovitine 45
– staurosporine 45
– suramin 45
– targets 48
– toyocamycin 45
cell culture 10
cell culture technique 82
cell cycle control 42
cell division cycle 42
cell surface proteins 16
central executioner 129, 130, 134
chromatin remodelling 166
classical proteomics 29
clustering 3
coactivators 166, 167
cultures as model systems 10

cyclin A1 51
cyclin A1(-/-) male mice 51
cyclin A2 51
cyclin B1 51
cyclin B1(-/-) mice 51
cyclin B2 51
cyclin B2(-/-) mice 51
cyclin D1 51
cyclin D2 51
cyclin D2(-/-) male mice 51
cyclin D3 51
cyclin I 51
cyclin-dependent kinases 39
– CDK1 42
– CDK2 42
– CDK4 42
– CDK5 43
– CDK5/p25 44
– CDK6 42
– CDK7 43
– CDK8 43
– CDK9 43
– neural and muscular functions 43
– transcription 43

database search 31, 36
defined serum-free medium 13
development 229, 231, 246, 248
diethylstilboestrol 177, 183
differentiation 231, 233
dihydrotestosterone (DHT) 89

efferent ducts 176, 177, 181, 183
epididymis 229, 230, 231
extracorporal storage 81
extrinsic pathway 130, 131, 134

Fas 145
Fas ligand 99
Fas receptor 99
fertility 70, 79, 81
fertility protection 81
follistatin 95, 98
FSH 88
FSH βsubunit gene 88
FSH receptors 92
– localization 92
functional proteomics 27, 30

GAK 50
gene arrays 1
gene expression 107, 115, 234
gene expression arrays 48
GeneChips 1
genetic manipulation 59, 60, 61, 71, 77
germ cells 60, 63, 69, 71, 78, 81, 197, 204, 206
germ cell apoptosis 141
germ cell movement 199, 205, 209
GnRH antagonist 177, 183
gonocytes 94

high-density oligonucleotide arrays 1
histone acetyl transferase 170
HSP70-2 50

ICSI 61, 66
immunogens 21
inhibin 98
inhibin/activin/follistatin system 98
intra-testicular concentrations 89
intrinsic pathway 131, 134

knockouts 149

Leukaemia inhibitory factor 95
Leydig cells 50, 99, 176, 180, 185, 187
LH 88
LIM domains 117, 120, 122
LXXLL motifs 167, 168

magnetic cell sorting 72
mass spectrometry 30, 31, 34
media optimization 15
microinjection 59, 63, 66
mitochondrial membrane permeabilization 134, 136
monoclonal antibody 17

neonatal rats hypothyroid 93
novel protein 28, 30, 36

oestrogen receptors 174
oestrogens 176, 177, 180, 185, 189
oncological patients 70, 82
organ culture 76

p53 144
p53/FasL double deficient mouse 151
paracrine regulation 199
PCTAIRE-1 50
PCTAIRE-3 50
peptide map 31
peptide sequence tag 34, 35
phosphorylation 107, 109, 111, 114
PITSLRE 44
platelet derived growth factor 95
precursor cells 18
progenitor cells 18
prostate 177, 181
protein kinases 39
"PSTAIRE" motif 41

radiation injury 150
5α-reductase 244–247
regulatory mechanisms 87
rete testis 177, 181

secretion 237
seminiferous tubules 60, 63
Sertoli cells 50, 92, 174, 189, 197, 199, 200, 204–207
– proliferation 93
signal transduction 107, 109, 115
spermatids 89
spermatogenesis 69, 71, 78, 79, 88, 90, 91, 92, 94, 108, 115, 118, 142, 199, 204, 205
– maintenance 91
spermatogonia 69–72, 78, 94
spermatozoa 66
spermiation 90
SRC1 167
stem cell factor 73, 77, 80, 97
stem cells 18, 69, 70, 76
steroid receptor coactivator 1 166

T receptors 92
– localization 92
techniques 9
testis 71, 78, 79, 81, 142, 174, 176, 181, 183, 189, 197, 203, 205
testis-specific co-activator 115
testosterone 88
TIF2 167
tools 9
transcription factors 107, 108, 112, 114, 117
transfection 70, 75, 77
transgenesis 59, 66
transgenic animals 59
transplantation 59, 63, 70, 76, 78–82
trinucleotide repeat 91

Ernst Schering Research Foundation Workshop

Editors: Günter Stock
Monika Lessl

Vol. 1 *(1991)*: Bioscience ⇋ Society – Workshop Report
Editors: D. J. Roy, B. E. Wynne, R. W. Old

Vol. 2 (1991): Round Table Discussion on Bioscience ⇋ Society
Editor: J. J. Cherfas

Vol. 3 (1991): Excitatory Amino Acids and Second Messenger Systems
Editors: V. I. Teichberg, L. Turski

Vol. 4 (1992): Spermatogenesis – Fertilization – Contraception
Editors: E. Nieschlag, U.-F. Habenicht

Vol. 5 (1992): Sex Steroids and the Cardiovascular System
Editors: P. Ramwell, G. Rubanyi, E. Schillinger

Vol. 6 (1993): Transgenic Animals as Model Systems for Human Diseases
Editors: E. F. Wagner, F. Theuring

Vol. 7 (1993): Basic Mechanisms Controlling Term and Preterm Birth
Editors: K. Chwalisz, R. E. Garfield

Vol. 8 (1994): Health Care 2010
Editors: C. Bezold, K. Knabner

Vol. 9 (1994): Sex Steroids and Bone
Editors: R. Ziegler, J. Pfeilschifter, M. Bräutigam

Vol. 10 (1994): Nongenotoxic Carcinogenesis
Editors: A. Cockburn, L. Smith

Vol. 11 (1994): Cell Culture in Pharmaceutical Research
Editors: N. E. Fusenig, H. Graf

Vol. 12 (1994): Interactions Between Adjuvants, Agrochemical and Target Organisms
Editors: P. J. Holloway, R. T. Rees, D. Stock

Vol. 13 (1994): Assessment of the Use of Single Cytochrome P450 Enzymes in Drug Research
Editors: M. R. Waterman, M. Hildebrand

Vol. 14 (1995): Apoptosis in Hormone-Dependent Cancers
Editors: M. Tenniswood, H. Michna

Vol. 15 (1995): Computer Aided Drug Design in Industrial Research
Editors: E. C. Herrmann, R. Franke

Vol. 16 (1995): Organ-Selective Actions of Steroid Hormones
Editors: D. T. Baird, G. Schütz, R. Krattenmacher

Vol. 17 (1996): Alzheimer's Disease
Editors: J.D. Turner, K. Beyreuther, F. Theuring

Vol. 18 (1997): The Endometrium as a Target for Contraception
Editors: H.M. Beier, M.J.K. Harper, K. Chwalisz

Vol. 19 (1997): EGF Receptor in Tumor Growth and Progression
Editors: R. B. Lichtner, R. N. Harkins

Vol. 20 (1997): Cellular Therapy
Editors: H. Wekerle, H. Graf, J.D. Turner

Vol. 21 (1997): Nitric Oxide, Cytochromes P 450, and Sexual Steroid Hormones
Editors: J.R. Lancaster, J.F. Parkinson

Vol. 22 (1997): Impact of Molecular Biology and New Technical Developments in Diagnostic Imaging
Editors: W. Semmler, M. Schwaiger

Vol. 23 (1998): Excitatory Amino Acids
Editors: P.H. Seeburg, I. Bresink, L. Turski

Vol. 24 (1998): Molecular Basis of Sex Hormone Receptor Function
Editors: H. Gronemeyer, U. Fuhrmann, K. Parczyk

Vol. 25 (1998): Novel Approaches to Treatment of Osteoporosis
Editors: R.G.G. Russell, T.M. Skerry, U. Kollenkirchen

Vol. 26 (1998): Recent Trends in Molecular Recognition
Editors: F. Diederich, H. Künzer

Vol. 27 (1998): Gene Therapy
Editors: R.E. Sobol, K.J. Scanlon, E. Nestaas

Vol. 28 (1999): Therapeutic Angiogenesis
Editors: J.A. Dormandy, W.P. Dole, G.M. Rubanyi

Vol. 29 (2000): Of Fish, Fly, Worm and Man
Editors: C. Nüsslein-Volhard, J. Krätzschmar

Vol. 30 (2000): Therapeutic Vaccination Strategies
Editors: H. Hennekes, W. Sterry, P. Walden

Vol. 31 (2000): Advances in Eicosanoid Reseach
Editors: C.N. Serhan, H.D. Perez

Supplement 1 (1994): Molecular and Cellular Endocrinology of the Testis
Editors: G. Verhoeven, U.-F. Habenicht

Supplement 2 (1997): Signal Transduction in Testicular Cells
Editors: V. Hansson, F. O. Levy, K. Taskén

Supplement 3 (1998): Testicular Function:
From Gene Expression to Genetic Manipulation
Editors: M. Stefanini et al.

Supplement 4 (2000): Hormone Replacement Therapy
and Osteoporosis
Editors: J. Kato, H. Minaguchi, Y. Nishino

Supplement 5 (1999): Interferon:
The Dawn of Recombinant Protein Drugs
Editors: J. Lindenmann, W.D. Schleuning

Supplement 6 (2000): Testis, Epididymis and Technologies
in the Year 2000
Editors: B. Jégou, C. Pineau, J. Saez

Kolibri 360 GmbH
Ferdinand-Jühlke-Straße 7
99095 Erfurt, Deutschland
produktsicherheit@kolibri360.de